Aristotle's Powers and Responsibility for Nature

Berner Reihe philosophischer Studien

Herausgegeben von

Prof. em. Dr. Andreas Graeser u.a., Universität Bern

Band 44

Diese Reihe versammelt Arbeiten, die im Umfeld des Instituts für Philosophie der Universität Bern entstanden und sich entsprechend der hier weiten Ausrichtung mit sehr unterschiedlichen Thematiken und Denkern verschiedener Observanz auseinandersetzen. Dabei orientieren sich die Autorinnen und Autoren an den Standards von Klarheit und Kohärenz und stellen ihre Beiträge in den Dienst der Sache.

PETER LANG

Bern · Berlin · Bruxelles · Frankfurt am Main · New York · Oxford · Wien

Armin Höll-Steier

Venture Capital

Fund Certification, Performance Prediction and Learnings from the Past

PETER LANG
Internationaler Verlag der Wissenschaften

Bibliographic Information published by the Deutsche Nationalbibliothek
The Deutsche Nationalbibliothek lists this publication in the Deutsche Nationalbibliografie; detailed bibliographic data is available in the internet at http://dnb.d-nb.de.

Zugl.: München, Univ., Diss., 2010

D 19
ISSN 0531-7339
ISBN 978-3-631-61834-9

www.peterlang.de

Stephan Millett

Aristotle's Powers and Responsibility for Nature

PETER LANG
Bern · Berlin · Bruxelles · Frankfurt am Main · New York · Oxford · Wien

Bibliographic information published by die Deutsche Nationalbibliothek
Die Deutsche Nationalbibliothek lists this publication in the Deutsche Nationalbibliografie; detailed bibliographic data is available on the Internet at ‹http://dnb.d-nb.de›.

British Library Cataloguing-in-Publication Data: A catalogue record for this book is available from The British Library, Great Britain

Library of Congress Cataloging-in-Publication Data

Millett, Stephan
Aristotle's powers and responsibility for nature / Stephan Millett.
p. cm. – (Berner Reihe philosophischer Studien, ISSN 1421-4903 ; Bd. 44)
Includes index.
ISBN 978-3-0343-0679-9
1. Aristotle. 2. Teleology. 3. Philosophy of nature. 4. Biology–Philosophy.
5. Metaphysics. 6. Environmental ethics. I. Title.
B491.T4M55 2011
113–dc23

2011033633

ISSN 1421-4903
ISBN 978-3-0343-0679-9

Hochfeldstrasse 32, CH-3012 Bern, Switzerland
info@peterlang.com, www.peterlang.com

Printed in Switzerland

CONTENTS

Introduction

Forms of life are dying out at an unprecedented rate. But what is life, what does it mean to be living and why does it matter? Does it matter if species die out? Does it matter if human domination of the planet creates an impoverished world that becomes less and less habitable? Does it matter that we behave 'responsibly'? Even if we answer 'yes' to each of these—and we should—why should we approach these issues through metaphysics as this book does? The simple answer is that if we revise our understanding of the way the world is we have to revise our ethical relationship to it. And understanding our ethical relationship is a necessary pre-condition for acting responsibly.

This book sets out to re-assess how we understand the way the world is, what life itself is, what value there may be in living things and what this value means for those capable of recognising it. These are issues argued over in various forms over many years, starting with the ancient Greeks and, although there is very little acknowledgment of this, some key elements of ancient Greek thinking have persisted into some of the most influential modern theories of environmental ethics. Whitehead may have suggested that European philosophy was a footnote to Plato, but it turns out that some very important parts of contemporary environmental philosophy may be a footnote to Aristotle. Aristotle can, after all, be considered the first biologist. But this book turns primarily to his metaphysics, and not to his biological works or his *Ethics*.

From a re-examination of Aristotle focusing on his idea of the motive force immanent in living things, it becomes clear that it is individual living organisms that are the locus of value and that individual living things are the prime objects of moral consideration. This also puts decline of species into a clear value framework: it matters every time a species becomes extinct. It matters because a unique kind of value and a unique store of value—common to all members of a species and instantiated in individuals—is gone, forever. But are all individual living things of equal value? Yes, and no. There may be some element of a thing's value that it has in equal parts with all other living things, but the value of an individual is also related at least in part to its level of biological complexity—so that more complex beings have greater value than less complex beings. This, however, does not fully describe the value of any individual, because no individual can live in isolation from others, whether of the same species or not. Each individual must live in an environment that it co-creates with others—something that Jakob von Uexküll demonstrated with his concept of *Umwelt* and which was later interpreted by Thomas Sebeok and others in terms of biosemiosis. In the process of co-creating its environment, each individual necessarily values at least some other living thing—but this is not a conscious valuing—if only because without the one the other will suffer and possibly die.

An approach to ethics based on living individuals is a biocentric ethic. And it is not only possible, but also highly informative and useful, to have a biocentric ethic based on at least some of the Aristotelian way of viewing the world. The biocentric ethic in this book requires moral agents to take responsibility for all living things simply because they are living. This creates an ethic that applies equally to the human and the non-human world, so making superfluous the notion of a purely 'environmental' ethic. In this ethic of responsibility for nature the most significant object of moral concern is an individual living thing and its own unique nature. And the correct attitude toward individual living things is an attitude of responsibility—in all its complex forms. In some respects contemporary environmental ethics has moved on from discussions of biocentrism and related metaphysical questions, but this book sets out to show that it has done so a little too quickly and that there is still much fertile ground to cover. The coverage of Aristotle and Spinoza in the first two chapters can be quite technical at times, but is there to demonstrate incontrovertibly the value of revisiting Aristotle in respect of environmental metaphysics and ethics. Some readers may feel inclined to skip these chapters, but should resist this—or at least come back to them—because concepts explained here recur throughout the book in a variety of forms.

There are environmental ethics that have urged humans to take responsibility for nature, but these have often assumed that 'nature'—whatever it is—is there simply for the use and benefit of humans and have made simplistic assumptions as to what 'nature' is. In short, they have not explored the metaphysical bases for their environmental ethics and have thereby limited the possibilities. There have also been ethics concerned with the welfare of non-humans that have taken metaphysics into account when determining what should be morally considerable and what should not. Chief among them is Bernard Rollin's pro-animal ethical theory which is based explicitly on an Aristotelian understanding of a living creature. The present work goes beyond Rollin in a number of ways.

This book explores in detail the question of what 'nature' is. This is a primarily metaphysical exploration, but one that also examines key aspects of modern biology—and establishes a clear relation between Aristotle's conception of 'nature' and other conceptions relevant to contemporary environmental philosophy. Aristotle to date has not figured as prominently in the environmental literature as he perhaps should, although he has been given a significant status in work on animal rights, through the work of Rollin, and more particularly through the work of Stephen R. L. Clark, and has had something of a revival in various forms of environmental virtue theory. However, outside of Rollin and Clark, and narrow uses of Aristotelian virtues, when Aristotle has appeared in an environmental context, it has generally been to receive what is colloquially known as 'bad press'. One of the aims of this book is to attempt an environmental rehabilitation of Aristotle—to show that an Aristotelian biological metaphysics can be coupled with an Aristotle-inspired moral philosophy to produce a biocentric environmental ethic. The initial focus, however, is not on ethics, but rather on biology and considerations of a biological metaphysics.

The book began life in a desire to examine claims that Humberto Maturana and Francisco Varela's concept of autopoiesis—a fringe player in the philosophy of biology, but one with an expanding sphere of influence—had dispensed with the need for any 'peculiar directing force' in explaining what it means to be a living thing. The term 'autopoiesis' was created in 1972 by Maturana and Varela and applied to single cells and multi-cellular organisms as part of an attempt to describe the necessary and sufficient conditions for the presence of life. One of the interesting claims made of autopoiesis is that its use can be extended beyond its origins in molecular biology and can be applied to such things as social systems and ecosystems. But, perhaps more important is the claim that the presence of autopoiesis is a defining characteristic of morally considerable entities: that an autopoietic unity is the paradigm case of moral considerability. This combination would make it useful not only for a biocentric ethic, but also for an ecocentric ethic. It turns out, however, that Maturana and Varela's theory is not fully coherent, that its claim to dispense with teleology in nature is deeply flawed and that use of autopoiesis should at the very least be restricted to the cellular level where it originated. It certainly should not be extended to cover social systems or ecosystems, but that is not to say that the ideas expressed in autopoiesis and the questions the concept tries to address are not important for environmental philosophy: they are. It is just that although ecosystems are an undoubtedly important element of environmental ethics they should not (and certainly not on the basis of any autopoietic character they might have) be considered individuals in their own right, as is claimed of them by some proponents of autopoiesis. It is better to see ecosystems as standing in a relationship of non-causal dependence to the biological individuals that comprise them—a relationship in which ecosystems have the character they do in large part because of the particular character of their component parts. The component parts of most interest here are living individuals, but working out just what makes something an individual is not an easy task. In examining the question of what an individual is—and how an individual maintains itself as an individual despite external influences and internal changes—two key concepts emerge: Aristotle's concept of *dunamis* and Spinoza's concept of *conatus*. There are several things of interest in these concepts. On closer examination, for example, it becomes clear that autopoiesis can be expressed in terms of *dūnamis* and that Spinoza's *conatus* both has a strong historical connection to *dūnamis* and also can be expressed in terms of it. Spinoza's concept of *conatus* (an internal impetus to maintain oneself in one's own being) is used extensively in deep ecology approaches to environmental ethics, notably Arne Naess's, and is a crucial component of Mathews' ecological metaphysics. For Stephen Clark there is a certain irony in this because Spinoza himself was adamantly anti-animal, holding not only that it was unnecessary to consider the interests and feelings of 'animals', but that it was actually wrong. A reassessment of *conatus* must, of necessity imply a reassessment of analyses that rely on the concept. So, although Naess's work is dealt with only in passing, if the interpretation of Spinoza's *conatus* outlined here holds then Naess also needs to be reconsidered. However, the connection made here between Aristotle and Spinoza is not entirely new: for example, Stephen Clark has referred to it in his work on animal rights. One of the results, however, of using it in the context of environmental metaphysics is that contemporary use of the concept of *conatus*, when examined closely, becomes, in fact, a concept much closer to Aristotle's *dūnamis* than to Spinoza's *conatus*. The close relationship between *dūnamis* and *conatus* by itself puts Aristotle firmly into the heart of environmental ethics, but the rehabilitation of Aristotle's reputation in environmental terms does not stop there. It turns out that other key contemporary theories in environmental ethics may also be read in Aristotelian terms and that the immanent purposiveness in all living things that Aristotle identified is a suitable foundation on which to build an ethic.

One of the most significant theories of environmental ethics that can be shown to be implicitly Aristotelian is in Paul Taylor's *Respect for Nature.* But, although Taylor's theory has been highly influential and is in most respects a cogent and well-developed approach, it suffers from having an underdeveloped concept of 'nature' and from flaws in the principles used to decide competing claims for moral consideration. His theory is based on an implicit Aristotelian teleology, but it turns out not to be fully-consistent. Freya Mathews' theory of the 'ecological self' is another significant theory with an implied debt to Aristotle: the central concept here of 'self-realization' being expressed in terms of what she claims is a Spinozist concept of *conatus*. But Mathews' concept of *conatus* turns out to be in all essential respects the same as *dūnamis*. Mathews' theory is also more cosmological than ethical: it seeks to establish a new cosmology as a foundation for ethics. But, although it provides a wide class of morally-considerable entities it offers little in terms of practical ethical guidance. A final theory, this time acknowledging an explicit debt to Aristotle's biological ontology, is the theory of responsibility put forward by Hans Jonas. The theory, as Jonas presents it, is thoroughly anthropocentric—which presents its own problems, but there is a more serious flaw: Jonas appends to his anthropocentrism an assumption that only a totalitarian and coercive politics is adequate to ensure that the duty of responsibility to living things is met. Such a politics is, however, not necessary particularly if moral agents take up responsibility and cultivate it as a virtue, this virtue of responsibility becoming the heart of a new environmental ethic. This virtue ethic differs from one such as Kant's, for example, in that, following Emmanual Lévinas, there is both an *a priori* obligation to ask what one might do to help an individual and a primary motive force for a moral agent to take responsibility for the welfare of an individual moral subject based in a *feeling* within the moral agent. The impetus behind moral agents taking responsibility for individual living things is a form of *love*—a love that is part of the very fabric of (biological) existence and one in which there is always already an obligation to help others.

Chapter 1 shows that Aristotle's biological teleology is important as a basis from which to understand moral considerability and places it at the heart of contemporary environmental metaphysics—and therefore at the heart of an ethic of responsibility for the environment. Discussion of biological teleology begins with Marjorie Grene for whom teleological concepts enter necessarily into philosophers' reflection about the data of science and perform a regulative function in the laboratory. In understanding Aristotle's concept of *telos* we must, however, take care to clarify misconceptions, a task which takes up the next part of the chapter. This includes a detailed examination of causation and analysis of key terms in Aristotle, particularly *hou heneka* (that for the sake of which), *eidos* (form or species) and *to ti ēn einai* (the-what-it-*is*-to-be). From there we move to an examination of *dūnamis* (immanent potential) and *phūsis* (nature) as key elements of natural teleology before dealing with the concepts of 'change', understanding of which is necessary to understand immanent (or directive) teleology, and 'self-movement'. The chapter concludes with a clarification of the nature of 'nature' (*phūsis*) and the internal principle of motion (*dūnamis*). Clarifying the relationship between these is important for the discussion in the next chapter which shows how the concept of *conatus* associated with Spinoza—and used in environmental philosophy—derives from an interpretation of *dūnamis*.

Conatus is a concept that attempts to explain how a thing—whether a rock, a tree, a butterfly or a mammal—continues to be what it is despite changes. The conatus of living things, however, is different from the conatus of non-living things, such as rocks and understanding what is special about the ability of living things to persist in their being is important to our understanding of the 'interests' we should take into account when dealing with living things. Demonstrating the relationship between Spinoza's *conatus* and Aristotle's *dūnamis* lays the ground for an ethics of responsibility that encompasses all living individuals. Chapter 2 is in three parts. The first deals with those aspects of Spinoza's intellectual heritage that influenced him in the formation of his concept of *conatus*. This begins by establishing a connection between Cicero's use of *conatus* and the Greek expression *hormē*, before examining how *conatus*, *hormē* and related concepts in Aquinas, Descartes and Hobbes influenced Spinoza. The second part compares Spinoza's and Aristotle's taxonomies of volition and develops a connection between some of

Aristotle's concepts and Spinoza's concept of *conatus*. This is a weaker connection than that demonstrated in the third section, but coverage of it helps show the degree of Spinoza's indebtedness to Aristotle. The third part compares Spinoza's concept of *conatus* and Aristotle's concept of change—particularly that part centring on his concept of *dūnamis*. Possession of a special kind of *dūnamis* sets living things apart from other objects in the world and is the basis of their autonomy.

The main issue in the first part of Chapter 3 is whether *individual* organisms are the only morally considerable entities and if they are, what this means for things such as ecosystems or communities. That individual organisms are morally considerable is intuitively satisfying, but to go beyond intuition we need to look at some of the claims made on behalf of ecosystems—such as the claim that ecosystems are what is known as autopoietic (self-creating) entities. But classifying ecosystems as autopoietic entities is shown to be flawed so discussion moves to the idea that ecosystem value supervenes on biological value. This supervenient relationship allows value to persist in ecosystems without imputing any ontological singularity to ecosystems or communities. A detailed coverage of both supervenience and ontological singularity is undertaken to clarify these difficult concepts before the discussion moves to causal 'powers' as these are implicit in Brennan's account of the features of a unified thing. From an Aristotelian perspective there is interaction and a relationship of reliance between the elements that make up an ecosystem, but the ecosystem or community itself has no *internal* principle of unity and is therefore not a source of value in the way an organism is. It is not an 'end-in-itself' with 'interests'. From ends-in-themselves the discussion moves to the idea of intrinsic value, another contested term in need of clarification, and then to the question of what an 'interest' is, what things have an interest and in what ways having interests is relevant to ethics. From there a discussion of intrinsic value theories in environmental philosophy leads to a discussion of so-called third-order autopoietic unities and the conclusion that there are no such unities and that ecosystems have value only to the extent that it supervenes on the value in individual organisms.

Chapter 4 focuses primarily on the biocentric environmental ethic that Paul Taylor sets out in his book *Respect for Nature*, with particular emphasis on the idea that individual organisms are teleological centres of activity. Taylor argues that we should demonstrate respect for nature, but does not explain sufficiently what ‘nature’ is that we might treat it with respect, nor does he clearly articulate what a living individual is, so both of these are addressed here in some detail. The question of what is an individual is complicated by the paradox of unity, something that requires careful explanation before it is possible to situate value in living individuals as entities with a good of their own and to move to the question of how to distinguish in value terms between different morally-considerable entities.

Chapter 5 builds on the previous chapters to show that an Aristotelian biological ontology is broad enough to be applicable to life-based ethics in general, but it counsels caution that we should not accept this applicability too hastily because there are important questions of metaphysics that need to be addressed—particularly the question of whether a holistic metaphysic is appropriate to a full understanding of value in and responsibility for the natural world. The chapter examines, in particular, Freya Mathews’ use of *conatus* and demonstrates that an ethic based on the concept of *conatus* may be expressed, without significant loss, in terms of an Aristotelian teleological biology and not be tied necessarily to a metaphysical position such as holism or monism. The chapter begins with Mathews’ cosmological approach then examines how she deals with the question of individuation before arguing that her use of *conatus*, which is the principle of individuation in her cosmology of substance monism, is closer to the Aristotelian archetype *dūnamis* than it is to Spinoza’s concept of *conatus*. From this it is clear that an Aristotelian teleological biology is a sufficient basis, or at least a sufficient ontological basis, for an environmental ethic.

Chapter 6 looks at three aspects of Aristotelian biological teleology that need in particular to be expanded and clarified. The first is the idea that, with the emergence of entities with a good-of-their-own, interest and value enters the world. The second is a claim that the relative complexity of entities is important, with more complex organisms having a stronger *prima facie* claim on moral agents than do less complex organisms. The third aspect is the notion that the obligation under which moral agents are placed is one of *responsibility*, where responsibility is an obligation that requires moral agents to *act* to protect the good of moral subjects (or moral patients) and where all living things are moral subjects. The question of how value enters the world presupposes an understanding of what value is, but value is an elusive concept, so what it might be is discussed at some length. Complexity is shown to be relevant to understanding the relative value of things in the world, through the correlated concepts ecological niche and interconnectedness and the idea of biosemiosis that emerged from biologist Jakob von Uexküll's work on *Umwelt*—a phenomenological explanation of how an organism engages with the world external to it. The increased capacity or potential of complex organisms to interact with their environment and the greater number of other organisms it can interact with is seen to give it an 'ontological niche' value as well as a 'semiotic' and an ecological niche value. From value and complexity the chapter moves to an explication of responsibility relying on the work of Emmanuel Lévinas and Hans Jonas, both of whom react to Heidegger's human-nature dualism. There follows a discussion of Jonas' notion of substantive (or forward-looking) responsibility and its applicability beyond the human, to the natural world.

Chapter 7 focuses on recognition of responsibility and what happens when this responsibility is recognised by a moral agent. It begins by describing how it is possible to move from a claim about the nature of things in the world to an ethical standpoint with regard to those things—making use of ideas from Holmes Rolston III, Emmanuel Lévinas, Martin Heidegger and Hans Jonas. The section following this deals with responsibility, virtue and feeling and the idea that moral agents are 'called' to behave well toward that which has value in and of itself. The 'call' is recognised—for Lévinas—in the 'epiphany of the face' an analogous understanding of which occurs when we encounter and engage with non-human living things. In this analogous understanding we engage semiotically with the natural world and in the first instance 'feel' a claim by a living thing and then respond to that feeling. The appropriate response is to take substantive (forward-looking) responsibility for a living thing and thereby cultivate responsibility as a virtue—a virtue that begins as a *feeling* in the moral agent and which, having arisen, is put into practice through the use of rational judgment. The feeling that *triggers* virtue is 'love', but love of a special sort. And it is this 'love' that allows a moral agent to recognize that he or she ought to take substantive responsibility for any living individual. The sort of love referred to here is much the same as an aspect of love described by Spinoza, for whom 'love is a joy,' 'accompanied by the idea of an external cause'. From 'love' as a trigger for responsibility, the chapter moves to a discussion of Hans Jonas' 'imperative' of responsibility, his human chauvinism and his totalitarian politics before concluding that neither human chauvinism nor a totalitarian politics is entailed in an ethic of responsibility. However, living things do demand something of us and if we deny the demand and do not act on it in a way that promotes the flourishing of individual living things we deny something essential about ourselves. Not to act, not to value through our actions that which has value in itself is the opposite of virtue.

PART I

THE HERITAGE OF ARISTOTLE

1. Aristotle's Biological Teleology

Introduction

There is no doubt that we recognise some things as morally important and that there are many reasons for this. But what is it that makes something morally important? There is a wide sense in which we can address this question and a narrow sense. In the widest sense we would look at human psychology, social connectedness, personal context and a host of other issues. And, we would focus on the thinking of moral agents. The focus of this book, however, is not to answer this question in its widest sense, but to address it in a narrower sense, a sense that looks to what might be inherent in an object that makes it morally considerable and which, in effect, announces a need to those around it. The story, if it can be called such, begins with Aristotle's notion of internal teleology and shows that this internal or, more correctly, immanent teleology is an undeniably important basis from which to understand moral considerability in the natural world. Once we understand this immanent teleology we can come to understand that it sits at the heart of contemporary environmental metaphysics—and ultimately at the heart of an ethic of responsibility for the environment.

Why begin with Aristotle? The are several reasons for this, some of which will become apparent in the coming chapters, but first among the reasons is that the teleology referred to is immanent (indwelling) in some living, concrete, identifiable individual and not in some grand process. This is satisfying at a common sense level and, as we shall see in the following chapters, firmly locates *value* in each living individual, so that there is something identifiable and tangible to which we, as moral agents, have an obligation. This also means that decisions that affect the natural world should be made with reference to the value of individual living things—one by one. But more of that later. Beginning with Aristotle's biology teleology is not merely to recognise a historical debt—as rich as this debt may be—to the first systematic biologist. More importantly, it provides the basis for a rich understanding of what it means for there to be purpose in nature. This is not a grand purpose, within a grand design, but undirected purposiveness within a living organism. Beginning with an account based in biological teleology allows us to understand that purposiveness in nature is a purposiveness that is not directed, but is directive; a purposiveness that inheres in living individuals of all sorts. Teleology, however, is often not well understood, so the early part of this chapter takes some time to explain it, with a special focus on the role teleology plays within modern biology.

The Concept of Teleology in the Philosophy of Biology

Dictionary definitions of teleology refer to 'the doctrine or study of ends or final causes', especially as it relates to evidence of design or purpose in nature[1] or, in shorthand, as the 'doctrine of final causes'[2]—an argument, derived from St. Thomas Aquinas' interpretation of Aristotle, for the universe being designed by an unmoved mover (God). But, like most dictionary definitions it is too great a simplification. Etymologically, teleology is the explanation (*logos*) of something in terms of the end (*telos*) it serves. *Logos* here is used in the sense of talk or discourse, although there are many other senses in which it has been and is used. *Telos*, as the explanation of something

1 W. Little, H.W. Fowler, J. Coulson, The Shorter Oxford English Dictionary on Historical Principles, Rev. & Ed. C. Onions, 3rd ed. Oxford: Clarendon, 1985, p.2255.

2 C. Onions ed., The Oxford Dictionary of English Etymology, Oxford: Clarendon, 1991, p. 908.

in terms of the ends it serves, has a long history that really began with first Plato's[3] and then Aristotle's notions of final causes, but the term *telos* itself is first recorded as being used by Homer (*Odyssey* 17.496) where it is translated as 'coming to pass, performance, consummation'.[4] Plato's phrase *haptoitu tou telous* in thc *Symposium* (211b) is translated by Liddell and Scott as 'the full realization, thc highest point, (the) ideal' of beauty, which carries with it the sense of 'to attain the end'. However, in Walter Hamilton's 1951 Penguin translation this phrase appears as that which is 'absolute, existing alone within itself, unique, eternal'. The *telos* is that which *is* absolutely, existing alone within itself. In the *Gorgias* (499e), *telos* appears in the phrase *telos einai hapasōn tōn praxeōn to agathon*— 'good is the object of all action'. The object here has the same sense as 'goal' or 'endpoint'. The notion of *good* being the object of all action is particularly important for Plato's pupil, Aristotle. For Aristotle, the good (*to agathon*) of or for an entity is identified with its *telos*—that is, what an entity aims at, whether consciously or not, must necessarily be a good for the entity.

It should be noted that Aristotle was primarily concerned with the *telos* of *individual living things*, as part of his general investigation of the phenomenon of change. His interest in explaining change in (especially) living things has given him a prominent place in the history of biology and his thinking is still a central component of contemporary discussions of teleology in biology. There has in modern times been some significant discussion of Aristotle's teleological biology. For example, in *The Idea of Nature* Collingwood[5] argued that there was a need to reinstate Aristotle's recognition of goals in nature. However, despite Collingwood's assertion that to some degree this reinstatement had already occurred there is ample evidence that Aristotle's sense of goals within nature

3 e.g. Phaedo 97c6-d3, e1-4; 986a6-b3; 99a7.

4 Henry George Liddell and Robert Scott, A Greek-English Lexicon, 9th ed. Revised Henry Stuart Jones and Roderick McKenzie, Oxford: Clarendon, 1990.

5 Robin George Collingwood, The Idea of Nature, Oxford: Clarendon, 1945.

has not been revived on any large scale, except within the philosophy of biology where Aristotle's biological ontology is frequently discussed. The Deep Ecology and New Age movements, in some at least of their various forms, may well be modern examples of belief in internal goals in nature, but none invokes the spirit of Aristotle. In fact, outside of work in environmental virtue ethics, Aristotle gets some bad press as he is regarded as holding that nature was merely a resource to be used for human benefit.[6]

Outside of the philosophy of biology teleological views of nature have tended to follow the Renaissance model in which any sense of purposiveness in organic nature is thought to be imposed from outside or to be reliant on consciousness. Within the philosophy of biology, however, teleology has been considered very differently, and treated extensively. To get an idea of what teleology is within biology, there is no better place to start than Marjorie Grene's discussion.

Teleological language, Grene says, 'persists in biology' but the language used is '*not*...the language of conscious purpose; as both Aristotle and Collingwood knew, the 'ends' of nature are not plans. Organic phenomena are directive, not directed.'[7] That is, there is something *inside* or *internal* to organisms that directs them to some *telos* or endpoint. They are not directed by some external force. In discussing the role of teleology in biology Grene is specifically *not* talking about teleology as i) conscious purpose, i.e. purposive behaviour or ii) the 'vast cosmological teleology which seems to some people inherent in the conception of evolution.'[8] She lists three

6 Derek Wall ed., Green History: a reader in environmental literature, philosophy and politics London: Routledge, 1994. p. 54. Wall says this is a position held by e.g. J. Donald Hughes, Ecology in Ancient Civilizations, Albuquerque: University of New Mexico Press, 1975.

7 Marjorie Grene, The Understanding of Nature, Dordrecht: D. Reidel, 1974, p.174.

8 Ibid. With respect to this, Wardy argues that: 'On my scheme, Aristotle retains his unique position in the history of natural philosophy by advocating teleology free of both design and cosmic subordination. (Robert Wardy, Aristotelian rainfall or the lore of averages', Phronesis Vol. 38, 1. p. 29)

general types of phenomena which seem to demand teleological language when we speak of them and six main ways to understand teleology. These three types of phenomena are: part-whole relations or structures found in morphology; means-end relations or structures found in physiology; and directed processes of the sort most obviously exemplified in individual development.[9] Reference to teleology can be characterized as 1) reflective, 2) regulative, 3) descriptive, 4) operational, 5) explanatory or 6) ontological. The first five of these relate to teleological *explanation* generally. The sixth pertains particularly to Aristotle's sense of teleology, however it is worthwhile briefly discussing each of these senses of teleology to get a complete sense of what teleology is in biology.

Teleological concepts enter into the philosopher's *reflection* about the data of science and in at least two senses also have a *regulative* function within the biological laboratory. The first is a fairly simple sense in which certain concepts may regulate a scientist's *choice* of data and the other is a more complex, *Kantian*, sense. Kant's regulative principle is a 'principle of reason which serves as a *rule*, postulating what we ought to do in the regress [toward empirical knowledge] but *not anticipating* what is present *in the object as it is in itself, prior to all regress.*'[10] That is, the principle of reason does not give us direct access to objects in the world, but does help us achieve a greater level of understanding of what objects in the world might be. Citing 1934 work by A.F. Baker,[11] Grene points out, with respect to the first sense of the regulative role of concepts, that a scientist studying, say, the development of a fungus, must already know

9 Grene, M. 'Time and Teleology', The Knower and the Known. London: Faber and Faber, 1966, p. 228-9

10 Immanuel Kant, Critique of Pure Reason, Trans. Norman Kemp Smith, London: Macmillan, 1992, A509, B537.

11 In her 1976 work 'Aristotle and Modern Biology' (in Grene and Mendelsohn (1976) eds., Topics in the Philosophy of Biology, Dordrecht: Reidel), Grene cites Baker's work in more detail as: A.F. Baker, Purpose and Natural Selection: a Defence of Teleology, Scientific Journal of the Royal College of Science 4, 1934, pp. 106-19.

something about the orderly development of the organism if the right 'facts' are to be selected for analysis. Both the interrelation of parts with reference to the whole, and the orderly development of the organism under investigation, must have been assumed before the 'right' facts could be selected for further investigation and analysis. For Grene, however, every question asked also contains a concept of 'that for the sake of which' the development is occurring, that is, a concept of its natural *telos.* For her, such considerations place teleological thinking in the position of at least a *regulative* principle (in the Kantian sense) at the beginning of biological research. A Kantian regulative principle cannot say anything about an object as it is in itself, it 'cannot tell us *what an object is*, but only *how the empirical regress is to be carried out* so as to arrive at the complete concept of the object...'[12] This regress is a 'task' (Kant A508, B536) i.e. an attempt to reach a 'maximum' of the series of conditions in a sensible world. The maximum is not given and the regress, or task of trying to reach a maximum (which might be seen as trying to reach perfect understanding), continues indefinitely (Kant is at pains to point out that this is an indefinite as opposed to an infinite regress). From this point of view, says Grene, *telos* is a reflective concept that acts as a signpost to the study of nature. Teleological concepts, therefore, are not merely meta-scientific, but are at the least regulative of the biologists' choice of data and the problems she chooses to examine and in this fashion enter into the scientists' reflection about the data of science. Grene is clear, however, that teleology should not be treated as *only* a regulative idea. She wants to know if there can be 'real processes with natural endpoints, real *telē* in nature.' In other words, she is concerned with the question of whether there are objective and not merely hypothetical explanations of the working of nature. This is an important matter on which there have been many attacks, and it is part of the intention of this chapter to demonstrate that there can be, in fact, objective (Aristotelian) *telē* in nature.

12 Kant, Critique of Pure Reason A510, B538.

Grene argues that there are *descriptive* and *operational* functions for teleological concepts in biology: descriptive because a biologist cannot proceed to describe what it is that she is analysing without referring to structures, uses, or achievements; and *operational* because teleological concepts operate within the procedures of biology in that they determine the way phenomena are described and the way experiments and subsequent analyses are conducted. Biologists describe phenomena in terms of existing structures and types. For example, Oppenheimer argues that in embryology the experimenter can succeed only if questions are put which the 'embryo can comprehend'.[13] Answers to the experimenter's questions come at the 'supracellular level' from 'the embryo alive'. In this way teleological concepts 'control not only the way the phenomena are described but the way experimentations with them and therefore the subsequent scientific analysis of them itself proceeds. We have teleological discourse, then, firmly within the *procedures* of biological science.'[14].

In addition to the reflective, regulative, descriptive and operational aspects of teleology in science, the application of information theory to biology suggests *explanation* also can be teleological. The transmission of the DNA code, for example can be considered as either a problem in information theory generating a finalistic explanation or as a causal explanation which would 'involve knowing how it can happen that these three groups of nucleotides attract this amino acid, what are the effective forces of attraction among the atomic groupings'.[15] Grene suggests that the explanation itself is teleological, and on two levels: the level of the message and the level of the particulars comprising the message. The message, as it is understood, cannot be reduced to its component particulars whereas the component particulars are explained by their role in the message. Grene makes her point as follows:

13 The idea of asking a question that the embryo can comprehend is revisited in Chapter 6 in the context of the Umwelt.

14 Grene, Understanding, p. 177.

15 Cited in Grene, Time and Teleology, p. 238.

> In interpreting developmental events in cybernetical terms...the scientist is explaining a lower level (the actual particulars) through reference to a higher (the message conveyed and 'understood')...such finalistic explanation is distinct in structure and explanatory power from the causal explanation which...is complementary to it.[16]

Finally, teleology may be treated as having an *ontological* importance which goes back to Aristotle and in which,

> any explanation that succeeds in explaining anything does so in virtue of our impression that we have achieved an understanding of something in the real world which we had previously found puzzling. If therefore teleological explanation is a genuine part of the knowledge of living things, it is so because *living things are not only apparently but genuinely telic* ... it is only the lingering authority of the machine analogy that prevents our admitting the truth of this statement.[17]

Grene notes that in order to assent to the 'telic character of living things, including ourselves, we shall need to complete a very fundamental revision of our basic ways of thinking. We need...a revisionist metaphysic to bring this about...'[18] She does not indicate what form such a revision might take, but the crucial point here is that this ontological sense of teleology in which individual organisms have an immanent purposiveness or *telos* has been largely ignored or forgotten. This immanent purposiveness can form the basis of a revised metaphysics, as we shall see in the following chapters.

16 Grene, Time and Teleology, p. 238.
17 Grene, Understanding, p.178. Emphasis added.
18 Ibid., p. 179.

Aristotle’s Teleology

The ontological sense of teleology goes back to Aristotle, but we should take care with Aristotle’s concept of *telos* as there are two main misconceptions. Aristotle’s *telos* is not a purpose or plan, but an intrinsic endpoint that will come about unless there is some form of external interference. This endpoint is ‘that for the sake of which’ things happen.[19] What will normally happen to a fertilized kookaburra’s egg? A baby kookaburra hatches from it. What will happen to the kookaburra? It will normally grow, mature and become an adult. The adult form is that for the sake of which a series of events happens; it is what will result unless something interferes.

The second misconception is that Aristotle is assumed to be interested in a single cosmic *telos*. The idea that a single cosmic *telos* was most important to Aristotle came through Judaeo-Christian appropriation of Aristotle’s ideas. In this appropriation an unmoved mover is necessary to end the infinite regress of causation. So, we ask, if parents cause their children to come into existence, who caused the parents to come into existence? If we keep asking this question we go back in time unless there is something that starts it all that does not itself need to be ‘started’ by something else. This is not the sort of teleology that is of interest here as the kinds of ‘ends’ that usually interest Aristotle are the determinate end points of particular processes within the natural world. The Aristotelian *telos* is the *intrinsic* endpoint in which a series normally culminates, that is, there is an *immanent* final cause in natural entities and it is this which is the ontological *telos*. This idea of immanent causality is evident throughout Aristotle’s biological works, where he invokes the concept of *telos* in the study of ontogenesis—the study of the coming into being and development of individual organisms. In his biology individual organisms have an internal *telos* in that they develop into,

19 Grene, Understanding, p.6.

or strive to develop into, mature specimens of the species of which they are a member. We need also to remember that it is not always easy to determine what parts of Aristotle's work should fall under the label of ontology as the verb 'to be' in Aristotle has many different uses *in different categories*' (e.g *Metaph.* 1017a 22-30).[20] Also, the being of an individual organism should not be considered in and of itself, as each organism is a representative of a species.

Cause

The species is in some sense a final 'cause' of the form taken by the individual, but when dealing with Aristotle it is not possible to talk of final causes in isolation from three other causes (*aitiai*) of which he speaks, particularly as Aristotle himself often linked and made inseparable two or more of his four causes. According to Aristotle, to explain why a certain change occurred is to 'cite some reason or factor in an explanatory 'because' clause'[21] and there are four main senses in which 'because' may answer the question 'why?'. These four ways of answering 'why' give rise to Aristotle's four causes: the material, formal, efficient (or moving) and final causes. Although these 'causes' are separable, in fact they seldom operate separately, with two or more generally coalescing to produce an effect. Understanding these causes is important to understanding the teleology immanent in living organisms.

(1) The *material cause*. Something with material substance must come from something with material substance: the statue is made from bronze, or marble, or some other *material* or matter. Both matter and form are *immanent* 'cause', as opposed to an external 'moving

20 See e.g. G.E.L. Owen, 'Aristotle on the snares of ontology', Renford Bambrough ed., New Essays on Plato and Aristotle, London: Routledge and Kegan Paul, 1965, pp. 70-71.

21 Andrew Woodfield, Teleology. Cambridge: Cambridge University Press, 1976, p. 4.

cause'[22] That matter and form are both immanent causes shows a bifurcation in Aristotle's ontology. The material cause is associated with matter alone, whereas there are three causes associated with Form—the formal cause, the efficient cause and the final cause. The ontology splitting along these lines does not, however, preclude any one of the four causes from operating with any of the others. These four causes can, in fact, operate singly or in combination with one or more of the others. When the causes combine, however, it is not always in line with the ontological distinction between matter and form. For example, the efficient and final causes may operate in concert, as might the efficient and formal.

(2) The *formal cause.* For something to exist it must take on a form. A particular four-legged carnivore, for instance, is a dog and not a cat because it has taken on the form or characteristics of its parents. A dog is what it is in large part because it has inherited the form of its progenitors: this heritable form 'causes' the animal to be a dog. It would be easy, but misleading, to equate the form of a creature with the *species* to which it belongs. In Aristotelian terms the form transmitted in procreation is not a species-form, but simply the form of the father—with a contribution from the mother. The individual living thing, with its capacity to produce another individual like itself, is the prime consideration in Aristotle's biological works and the notion of species is simply a generalization over individuals functionally organized for 'the better'.[23] The formal cause is associated most closely with *to ti ēn einai*—an important term most commonly translated as 'the-what-it-*is*-to-be'—which will be explored shortly.

(3) The *moving (or efficient) cause.* There must also be something that sets movement or change in motion or which stops it. This is the moving or efficient cause. An object at rest has the potential to move

22 Aristotle, Physics, transl. Wicksteed and Cornford, London: Wm. Heinemann, 1929 (Loeb edition), 194b25, translator's note 'a'.

23 Allan Gotthelf, 'Introduction', A. Gotthelf ed. Aristotle on Nature and Living Things, Pittsburgh: Mathesis, 1985; Bristol: Bristol Classical Press, 1985. p. xv.

but cannot move unless acted upon by something that is itself in motion. The agent of movement may be internal or external. Blair, for example, argues that Aristotle's use of *energeia* and *entelecheia* implies that for living entities there is a moving cause immanent to each individual.[24] Blair's argument treats the 'en-' prefix in these two terms as roughly equivalent to the English word 'internal', giving us 'that which has its source of movement internal to it' for *energeia* and 'that which has its end in itself' for *entelecheia.* Internal, or immanent, efficient causation is a key element in Aristotle's ontology but it should be made clear that there are also external efficient causes: a person, for example, can fall over because he or she is pushed (an external moving cause) or fall over by choice (an internal moving cause).

(4) The *final cause* is *hou heneka*, the end or purpose, 'that for the sake of which' a thing is done—and the cause most often associated with teleology. The final cause of an artefact is generally its intended function, but natural things are different. For Aristotle there is, *internal* to all natural things, a 'that for the sake of which' it exists which cannot be explained in terms of function.

Three Key Aristotelian Concepts

Teleology in Aristotle is more than simply a matter of equating *telos* with *hou heneka*, because all of the causes are involved. To get a better idea of teleology in Aristotle, in particular, and biology in general, the so-called final cause (*hou heneka*) needs to be examined together with two other key Aristotelian concepts: *eidos* and *to ti ēn einai.* These three concepts relate to what are commonly known as the

24 George Blair, Energeia and Entelecheia: 'Act' in Aristotle, Ottowa: University of Ottowa Press, 1992.

'final' and 'formal' causes, and although to get a complete picture it is necessary to deal with the terms separately, it is not really possible to separate them fully because each relies on the other in some degree for its meaning.'

There is considerable debate over the correct translation of *to ti ēn einai* which has often and misleadingly been translated as 'essence'. If Joseph Owens' analysis is correct, then 'the-what-it-*is*-to-be' is the best translation.[25]. Owens shows that the phrase derives from *to einai* plus the dative (the being of 'x'). For Owens *to* = 'the (answer to the question)', ti = 'what', ēn = 'was (all along), *einai* = '(the) being of something'. However, Burger, following Benardete, translates this as 'the-what-it-was-to-be',[26] the Benardete translation picking up on the combination of the imperfect (*ēn*) and infinitive (*einai*), tenses of the verb 'to be' used in the Greek phrase. In a translator's note to the Loeb edition of *Physics* (194b28) Cornford expresses *tou ti* ēn *einai* as the 'being-what-it-had-to-be', a translation that indicates that the thing could not have been otherwise than it is. Here, Cornford says 'that is to say, it [the thing in question] must have actually 'arrived' and realized its 'being-what-it-had-to-be'.' In other words, in one sense at least, it has reached an end that was immanent, or potential in it—for example, the lion cub is potentially the adult lion, the child is potentially the adult human, and so on. Burger claims the use of the imperfect verb in *to ti ēn einai* implies that 'the determination of what a concrete whole '*was* to be' seems to involve a retrospective reflection' and the identity of the concrete whole 'seems to be that of a potentiality fully actualized.'[27] In other words, the concrete whole had the (immanent) potential to be what it has become. This immanent potential is at the heart of an Aristotelian teleology. But the term *to ti ēn einai* does not of itself imply or suggest teleology, and the use of

25 Joseph Owens, The Doctrine of Being in the Aristotelian Metaphysics, Toronto, 3rd ed, 1978, appendix.

26 Ronna Burger, Is each thing the same as its essence?: On Metaphysics Z. 6-11, Review of Metaphysics 41 (September 1987): 53-76

27 Ibid., p. 71.

the 'philosophical imperfect' occurs also in Plato, for example, where Socrates and others use the imperfect to indicate that they are referring to the truth (whatever it may be) about something which was previously brought into question. The truth is what the thing *was* 'all along'.

At least part of the truth of a thing is the immanent *telos* or endpoint toward which it will ordinarily progress: there is *always* in nature an ordered passage to a definite endpoint—such as the robin that lays an egg which hatches and matures into an adult. This endpoint, or goal, need not be a conscious purpose but must be the endpoint of some *entity's* becoming. That is, there is an objective endpoint or objective teleology operating. In Grene's view, the Aristotelian concept of *telos* is intelligible and useful only to the extent that it refers to something already in existence. Therefore, she says, something like evolution cannot be teleological in this sense because it is not dealing with an existing entity, it is dealing with classes or species of individuals, rather than with individuals *per se*. It cannot be teleological, it can only appear so, and therefore can only *appear* to have an endpoint. Evolution gives an impression of being teleological because it deals with a process of adaptation: an observer may recognize that there is a process of adaptation occurring over a number of successive generations and conclude that the last observed adaptive change is one toward which the other changes have been heading, or the observer may conclude that the goal of the observed changes is something that has yet to come about. The process of adaptation thus cannot be teleological in the Aristotelian sense because it deals with a process, not with an already-existing entity. There is, however, a sense in which Aristotelian teleology does admit of processes. The explanation of prior stages in terms of subsequent results implies that Aristotle's teleology does involve explanation of processes toward goals that have not yet been realized. Gill, for example, argues that the form, as an active cause, accounts for the continuity of processes.[28]

In order to further the discussion of *to ti ēn einai* we need now briefly to examine the concept of *eidos* (as *form* or *species*). The concept of *eidos* is one avoided by modern biologists, perhaps because Aristotle's position that species are eternal and constant has been overtaken by Darwinian evolution. But the concept of *eidos*—as form or species—has at the least heuristic value in that it can help us to locate a real process or structure in nature.[29] In other words, there *is* an endpoint for the development of each organic phenomenon: the apple sapling will ordinarily become an apple tree; the fledgling robin will ordinarily become an adult robin. And mature specimens do exist as things in the world, as exemplars of that which members of a species will ordinarily become. The *eidos*, or form, is not simply an empty, extension-less ideal. For Aristotle, Grene says,

> ...form exhibits its *ontological* foundation in the 'being-what-it-is' [*to ti ēn einai*] of each kind of thing....The *ti ēn einai* is expressly what definition is about, both its target and its presupposition. Indeed without the *ti ēn einai*, as the referent of definition, the real being-what-it-is of this kind of thing which the defining phrase designates, the demonstrations that follow on definition would be simply 'hypothetico-deductive' in the sense of positivism or phenomenalism. They would not be rooted, as Aristotelian science was certainly meant to be, in the natures of the things themselves, and in our understanding of these natures.[30]

There is an ontological foundation to Aristotelian science. And the ontology is primarily a biological ontology—firmly rooted in real, existing, living phenomena.[31]

28 Mary Louise Gill, Aristotle on Substance: The Paradox of Unity, Princeton, NJ: Princeton University Press, 1989. Ch. 6 and 7.

29 Grene, Aristotle and modern biology, p. 20.

30 Ibid., p. 31.

31 Eduard Zeller, Outlines of the History of Greek Philosophy, 13th ed., New York: Dover, 1980, p.176. The point is made also by others, e.g. Freya Mathews, The Ecological Self, London: Routledge, 1991, p. 108. It is worth noting also the

In modern biology, one of the defining characteristics of a living thing is the capacity for autonomous movement—whether this movement is something like growth or more like walking. In Aristotle's primarily biological ontology living things have an internal capacity for change (or movement)—including the capacity (*dūnamis*) for self-change. For example, at *De Anima* II.4 Aristotle indicates that this internal capacity is an active capacity linked to the soul (*psychē*) in such a way that the soul is the form of the living being, the first efficient cause of nourishment and growth, a set of active capacities (*dūnameis*) that directs life activities. It is the soul, as a set of active potentialities or internal form, that determines the goal of change for autonomous natural entities. In an Aristotelian ontology, it is the soul that determines a particular identity and it is by virtue of the active potential that is within an organism that this or that change begins and ends. The active potential (*dūnamis*) can perhaps best be seen as a set of 'operating rules' for complex tasks.[32] The distinction between active and passive *dūnameis* is an important one that is discussed in detail in the next chapter.

Finally, there is *hou heneka.* The *hou heneka* for Aristotle is 'that for the sake of which' a thing is done. But, as Gotthelf demonstrates, nowhere in the Aristotle corpus is there a detailed analysis of what it means for a thing to be, or come to be, for the sake of something.[33] We can be pretty certain, however, that Aristotle took the end (*to telos*) to be included in his catalogue of causes and we can be very sure that the so-called 'final causes' are an essential part of Aristotelian teleology. Moreover, Aristotle's account of teleology in general is derived from his analysis of teleology in natural products (organisms) and demonstrates a preoccupation with 'that for the sake of which' in seeking an explanation of organic development. [34]

---

work of David Balme on this, for it is due to him that a clear correlation has been established between Aristotle's biology and his metaphysics.

32 Gill, Aristotle on Substance., pp. 200-201.

33 Gotthelf, Final Causality, p. 226.

34 Ibid., p. 230, n.8. On this point see also George Blair, Energeia and Entelecheia.

Aristotle's natural teleology is a central element of his thought and can be described in terms of his thesis that the development, structure and functioning of a living organism is for the sake of something. This natural teleology is drawn *from* his observations of nature and is not something brought *to* nature. Aristotle thought that biology could not be reduced to what are today called chemistry and physics and that organic development 'involves the actualization of an irreducible potential for form.' This irreducible potential is the heart of 'being for the sake of something.' That is, an immanent potential (*dūnamis*) is necessary for an organism to exist 'for the sake of something'—which is to say, that existing is a process *for the sake of* its end.[35] For Aristotle, there are goals in nature that originate within the entity having the goal—and where no external causal agent is needed. Teleology is immanent (or indwelling) in living things and no external agency is the source of what a thing is to be. This should be distinguished from the situation for artefacts, as Aristotle makes plain in *De Anima* II.1, where the active capacity for an artefact is *external* to the artefact.

That for the sake of which a thing is done is, for Aristotle, a good. That is, a goal for an entity is something that is good for that entity and recognizing the equivalence of the good for an entity and any goal it might have (consciously or not) is essential to a proper understanding of Aristotle's natural teleology.[36] The idea that the notions of 'good' and 'goal' for an entity are equivalent is something that Aristotle inherited from Plato[37] and becomes particularly important if we choose to take the view—spelled out in later chapters—that moral agents have the capacity to recognize the good that is immanent in living things and that, having recognized a good, they are under an obligation to foster that good. But what connection

35 Gotthelf, Final Causality, p. 254.

36 John Cooper, Aristotle on Natural Teleology, M. Schofield and M. Nussbaum eds., Language and Logos, Cambridge: Cambridge University Press, 1982, p. 197n.

37 Phaedo 97C6-D3, E1-4, 986A6-B3, 99A7-C7

is there between recognition of a good and any attempt to explain an event? The connection is simply, according to Aristotle, that all (biological) teleological explanations are claims that something happened *because it is good.* One example of a teleological explanation of this sort is the explanation of the fact that a living thing has parts in virtue of the good that these parts do for the living thing.

The *hou heneka* or 'final cause', is closely linked to the 'formal cause' (*to ti ēn einai*). That is, 'that-for-the-sake-of-which' a change takes place in an entity is linked to 'the-what-it-is-to-be' and both of these are linked to the efficient or moving cause, forming a trio of causes.[38] However, from the frequent grouping together of final, formal and efficient causes, and their distinction from the material cause, it should not be inferred that there is always a clear separation between matter and form. Aristotle certainly mixes and matches the four causes in a variety of applications, but not always to effect a split along formal/material lines. In the *Physics* the efficient, or moving, cause, for example, may be connected to the formal and final causes, but in the biological works in particular it is associated with the material cause. Aristotle uses a number of combinations, for example: efficient and final cause; formal and final; efficient and material; and formal and material. But the conjunction of formal, final and efficient causes is not necessary merely for a *new* living thing to come into being, as Cooper suggests. The continued operation of all of these causes *within* a living individual is also necessary for the living individual itself to *continue* to exist. These causes are immanent in and definitive of a living being.

Final causes are not normally external causes and Aristotle's teleology is one that is immanent in nature. To confuse immanent teleology with external teleology 'is like confusing "God made the animal move" with "God made the animal in such a way that it could

38 Physics 198a24-27. Trans. Cornford and Wicksteed. There is, of course, in Aristotle something which moves without itself being moved e.g. at Metaph. 1072a25-b5.

move".'[39] Immanent goals are determined from within the entity, consciously or unconsciously, with or without understanding. For Aristotle an entity will strive to become a true representative of a natural kind, but despite suggestions that Aristotle's view has much in common with a modern cybernetic, or systems view of the world,[40] Aristotle's world is not a self maintaining system. Aristotle is a steady-state theorist, his world is made up of a finite and fixed number of natural kinds and depends upon a Prime Mover as ultimate final cause.[41] The Aristotelian view of teleology in organisms should not be confused with the view that an end or purpose is something ascribed to a phenomenon from a position outside the phenomenon. For Aristotle, teleology is built in to the living fabric of organisms. It is not something we (humans) *bring* to an organism with our descriptions of it. That is, it is not merely part of a discourse about living systems, but is something necessary for the understanding of living organization. Natural teleology is not about purpose in a 'directed' sense: it is about purpose in a 'directive' sense. 'Directed' here implies that the direction or purpose of an entity is determined *outside* the entity whereas 'directive' implies that the direction or purpose is determined from *within* the entity. *Telos* for Aristotle is (in part at least) an attribute of the living system (organism) itself, that is, it is internal to the system. In anthropological terms, Aristotle's view might be seen as 'emic', that is, a view from within the system rather than an 'etic' view—a view from outside the (living) system. If we were to take a 'syst-etic' view of teleology—as something external to the object of study (as proponents of autopoiesis do) we would be dispensing with a long tradition of inquiry into the nature of teleology within science, and within biological science in particular, where teleology is taken as indeed something internal to living things. Taking a syst-etic view would mean not asking questions that the

39 Woodfield, Teleology, p. 11.

40 Ibid., p. 15.

41 David Sedley, Is Aristotle's Teleology Anthropocentric?, Phronesis 36.2, 1991, p. 186.

organism 'can comprehend'. Moreover, to understand what it might mean to ask questions which the organism can comprehend we also have to recognise that organisms engage semiotically with their world, their semiotic processes being dependent on their unique set of sensory apparatuses—an idea explored in Chapter 6.

Aristotle's 'Powers' and Change

To understand immanent (or directive) teleology we should also come to terms with the concept of change as Aristotle is concerned, especially in the *Physics,* with explaining the *change* or *movement* that occurs in 'nature'. In fact, for him change and 'nature' are inseparable. As he says

> nature (*phūsis*) is the distinctive form or quality of such things as have within themselves a principle of motion, such form or characteristic property not being separable from the things themselves, save conceptually.[42]

In reading this, however, we need to remember that Aristotle made use of *phūsis* in two senses: as the sum of all essential powers *and* as an innate urge.[43] This innate urge, or internal principle of motion, may be active or passive and when it is an active principle it may act on another entity or act on the entity possessing the active principle. The 'powers' are Aristotle's *dūnameis*, of which there are two distinct levels[44] a general level where *dūnamis* is 'a principle (or source) of motion in something other than the thing moved or in the same thing *qua* other,' and a second level where powers are present in things not

42 Aristotle, Physics II, i; 193b 3 (Loeb edition).
43 Alexander P. Mourelatos, Aristotle's 'powers' and modern empiricism, Ratio, Vol. 9, 1967, No. 2, pp. 97-104.
44 Ibid.

accidentally, as they would be, for example, in art, but essentially. This second level contains a smaller group of powers that constitutes the essence of a thing, its nature or *phūsis*. As Mourelatos explains, there is also included in this set of essential or natural powers

> … a power which involves no reference to specific *other* things, a power which is intrinsic to the thing ... a power which constitutes its *phūsis* in a most intimate and characteristic way. *Phūsis* in this special sense is 'a principle of motion and rest in respect of place, growth and decrease, or alteration (*Phys.* II, 1 192b13); it is an 'innate urge (*hormē*) of change' (192b18)...So understood, *phūsis* is neither an active nor a passive power; it is...a reflexive power, a power of self-development.'[45]

If living unities *actively* maintain their identity (i.e. maintain it themselves), then there is some *internal motive force* or principle that allows them to do this. The argument for this is simple: if the motive force were not internal to living unities, then it would not be they who were doing the maintaining.

Mourelatos uses a shorthand to differentiate two sorts of 'phūsis': for 'phūsis' as the sum of all essential powers he uses *phūsis* (italicized) and for 'phūsis' in the special sense of innate urge he uses PHŪSIS (capitalized). Using this shorthand he further clarifies the distinction:

> ...the active or passive powers which belong to the *phūsis* of a thing is what the thing will do *if* [a 'trigger' is activated]. But its PHŪSIS is what it will do *unless*... The PHŪSIS of a thing is what the thing will do normally, when there are no external hindrances...[46]

The powers (what Aristotle called *dūnameis*) can be either passive or active. For there to be movement of any sort in an object, it must have the potential (*dūnamis*) to move or be moved. To be moved an object

45 Ibid., p. 99-100. The link between hormē and conatus will be examined in the next chapter.

46 Ibid., p. 100.

needs the passive potential to be moved if it is acted upon by some force. For an object to move anything else it must also have an active *dūnamis*. The passive potential (*dūnamis*) is what the thing will do '*if*' an internal trigger is activated. This passive *dūnamis* is part of the being of the entity and is neither added to it nor separable from it.

Self-movement

For there to be the movement characteristic of living things there must be present a passive potential acted upon by an internal motive force. For Aristotle (*Physics* VIII.4, 255a5ff), self-movement, conceived of in the sense of *kinēsis* or all-embracing change, is a property only of living things. Despite the fact that at one point Aristotle modifies his doctrine of self-movement (at *Physics* 259b1-20) by noting that much of what appears to be self-motivation for living things is in fact a response to external stimuli, there is nothing in the relevant passage of the *Physics* to suggest that the motion of the living thing which constitutes its PHŪSIS, its orderly growth toward the realization of its peculiar form, is externally induced.[47] That is, in the Aristotelian scheme, the movement constituted by the orderly growth toward the realization of [an organism's] peculiar form is internally generated. Movement of the organism itself (such as walking, slithering, breathing) is also internally generated, but may be said to be the result of an active *dūnamis* acting on the organism's own passive *dūnamis*.

That an organism can have an active *dūnamis* acting on a passive *dūnamis* comes about because when *phūsis* is taken as an innate urge (PHŪSIS) it becomes, in effect, a species of the genus *dūnamis*—a claim supported by Gill's reading of Aristotle.[48] In the passage where she notes that *phūsis* is a species of *dūnamis*, Gill makes clear that, for Aristotle (in *Metaphysics* Θ.8), there is more than one type of motive

47 Ibid., p. 101n.

48 Gill, Aristotle on Substance, p. 218.

principle (*dūnamis*) in living things. There are *passive* motive principles, which enable an entity to be moved by a motive force, and there are *active* motive principles which enable an entity to do the moving. But there are two sorts of active motive principle: there is an active motive principle that causes change in another, and an active motive principle within a living thing that causes change in the thing itself. In *Metaph.* Θ.8 Aristotle says:

> 'I mean by *dūnamis* not only the one that has been defined, [in Θ.1] which is called an active principle of motion in another thing or *qua* other, but generally every active principle of change and rest. For nature (*phūsis*) is also in the same genus as *dūnamis*; for it is an active principle of change but not in another thing but in the thing itself *qua* itself.'(1049b5-10)[49]

Gill notes that Aristotle in *Metaph.* Θ.8 sees an entity's nature (*phūsis*) as an active *dūnamis* that is a source of change in the thing itself *qua* itself.

It is worth noting here that an active motive principle will not 'cause' a change in an entity (be it itself or another) without the entity that is acted upon possessing a corresponding passive motive principle. Change can be 'caused' in a living entity only because the possibility of change is always already 'built-in' to the entity undergoing change. That is, for an entity to react to an external stimulus it must already have within it a passive motive force (i.e. a passive *dūnamis*) that determines how it will (re)act. When a (living) entity changes *itself*, an active *internal* motive force joins with a passive internal motive force to allow the change. When a living thing maintains itself, as itself, it is its active *dūnamis* acting on its passive *dūnamis*.

So, how does an entity which is not an unanalyzable whole maintain itself, as itself, despite changes?[50] Part of this question has

49 Translated, Gill, Aristotle on Substance, p. 218.

50 Understanding this is important in understanding what we should count as being morally considerable, an issue discussed in Part II.

just been answered, but only part. Gill argues that Aristotle has two models to account for the relationship between potentiality (*dūnamis*) and actuality (*energeia/entelecheia*) and that it is the second of these models which explains how a composite persists as an individual, despite changes, once it has been generated. Gill, like Blair, accepts that *energeia* and *entelecheia* refer to the same basic concept in Aristotle. These terms—and how they are related—will be explained in the next chapter.

The first potentiality-actuality model includes an active and a passive potentiality and has two main actualities: a motion and a product. The second potentiality-actuality model has the same characteristics as the first model but with the addition of an agent and a patient—that which has a passive potentiality acted upon by the agent—and a different concept of motion. In the second model the agent acts and the patient suffers (feels the effect of an act) in respect of the same properties. When the agent acts it is in terms of a motion (*kinēsis*) that is not a change (*metabolē*) but an activity (*energeia*). The importance of this second potentiality-actuality model becomes obvious when the agent acts on itself i.e., when there is internally-generated activity. Gill explains that on the second model a product that has already been generated (i.e. a living thing) is kept as that product through an internal activity which is, in effect the result of an active and passive *dūnamis*.[51] She says the structure of the second model mirrors the first but with 'the single modification, that the agent act[ing] on itself *qua* itself, yields a system entirely different...'[52] She uses the example of an axe and an axe-maker. The axe-maker has an active potential and the materials also have an active potential. But the craftsman and the materials are *unlike*. The completed axe, by contrast, requires an axe-user 'who has an active capacity for the same activity for which the axe has the passive capacity'.[53] That is, like needs to act on like. The axe-user does not change the nature of the

51 Gill, Aristotle on Substance, p. 214.

52 Ibid., p. 218.

53 Ibid., p. 219.

axe when he uses it to chop, but merely 'enables the axe to manifest its full being in the activity for which it was made: chopping.' The activity of chopping with an axe is a joint expression of 'an agent and a patient whose active and passive *dūnameis* are like.'[54] There is, however, a fundamental difference between axe and axe-user. Artefacts such as an axe require a user, but 'in living things, user and implement are the same individual'[55]. This can be seen at *Metaphysics* Θ. 8, where Aristotle says that a nature (*phūsis*) is an active principle of change in the thing itself. In a living thing, nature is an active capacity (*dūnamis*) for living which uses the living thing's body as an implement. In effect the organism acts on itself, via its soul, in the activity of living.

Gill notes earlier that autonomous entities (such as living things) rely on themselves to realize their own capacities and also rely on themselves to persist as living entities. Both the realization of capacities (potentials) and the persistence of identity for a unity are types of change in Aristotle's scheme. But although there is an internally operating cause, on the Aristotelian view, it should be remembered that the cause is not exclusively internal—there are also external causes operating. However, what a thing does in response to external forces is determined by what it is—both its matter (what it is made of) and its form (what kind of thing it is).[56] The particular reaction of any unity to an outside influence is at least partly determined by the *materiality* of the unity. At the same time, there is *internal* to every unity a principle of organization, a principle which defines how the various components of an entity will be interrelated so that its identity will be maintained. In Chapter 6 the idea that an organism engages with and interprets its world through a process of biosemiosis will be explored. We can see here what might be viewed as an ontological basis for this semiosis.

54 Ibid., p. 219.
55 Ibid., p.219.
56 Cooper, 'Natural Teleology', p. 199.

The Nature of *Phūsis*

In Aristotle's system all living things have two natures, a formal and a material nature. Each of these natures acts, as it were, internally (although not exclusively so). At *Physics* II i Aristotle says,

> ...one way of regarding 'nature' (*phūsis*) [is] as the ultimately underlying material (*prōtē hylē*)) of all things that have in themselves the principle of movement and change. But from another point of view we may think of the nature of a thing as residing rather in its form (*morphē*), that is to say in the 'kind' (*eidos*) of thing it is by definition.[57]

But it would be a mistake to think that 'nature' (*phūsis*) can be investigated exclusively by reference just to its material or just to its form. Aristotle makes this plain a little later at *Physics* II iii:

> since two sorts of thing are called nature, the form and the matter, we must investigate its objects...neither independently of matter nor in terms of matter only.[58]

In all of this it is important to recognize that the English word 'nature' does not adequately translate the Greek *phūsis*. Heidegger, for instance, in his *Introduction to Metaphysics,* contends that:

57 Physics 193a 28-31, transl. Wicksteed

58 Physics 194a12-15 transl. Hardie and Gaye in Jonathan Barnes ed., The Complete Works of Aristotle, Princeton: Princeton University Press/Bollingen, 1984, p.331. Wicksteed and Cornford translate this passage as 'Since 'nature' is used ambiguously, either for the form or for the matter, Nature, as we have seen, can be regarded from two points of view...it can neither be isolated from the material subject in which it exists, nor is it constituted by it.'

> ...the late concepts of 'nature' must be set aside: *phūsis* means the emerging and arising, the spontaneous unfolding that lingers. In this power rest and motion are opened out of original unity.[59]

In his lectures on Aristotle's *phūsis* Heidegger notes that 'Nature becomes the word for Being'[60] and that the key issue in Aristotle's *Physics* is that of 'defining the Being of movement [*kinēsis*]'.[61] Heidegger says that the word 'nature' derives from the Roman translation of *phūsis* as *natura* and that '*natura* comes from *nasci*, to be born, to originate...' as in the Greek root 'gen-' (in e.g. the English word 'genesis'). *Natura,* for Heidegger, means 'that which lets something originate from itself'.'[62] Although Heidegger's predilection for sometimes obscure etymology is well-known, even notorious, Cornford also notes that 'English cannot reproduce the ambiguity of *phūsis* which (like the French *la nature*) may mean 'the nature' (of any natural thing) or 'Nature' collectively, the sum total of such natures.'[63]

Aristotle spends Book II chapter 1 of the *Physics* detailing the different senses of *phūsis*. Here he notes that the one common feature that characterizes things constituted by 'nature' is that

59 Martin Heidegger, An Introduction to Metaphysics, transl. Ralph Manheim, New Haven: Yale University Press, 1959, p. 61.

60 Martin Heidegger, On the being and conception of PHUSIS in Aristotle's Physics B, 1, transl. Thomas J. Sheehan, Man World, Vol. 9, 1976, p. 222. 'Being' in this work is a translation of das Sein. Heidegger was insistent that das Wesen and its forms not be translated as essence.

61 Ibid., p. 225.

62 Ibid., p. 221.

63 Aristotle, Physics (Loeb edition). Footnote at 192b 21. See also in the next chapter the distinction between *Natura naturans* and *Natura naturata.*

> each of them has within itself a principle of motion [*archēn kinēseōs*] and of stationariness (in respect of place, or of growth and decrease, or by way of alteration).[64]

The key element of this is the notion of an *internal* principle of motion (as opposed to an external cause). If this internal principle of motion is interpreted as a *dūnamis* then Cornford's observation about the ambiguity between different senses of *phūsis* is much the same as Mourelatos's distinction between *phūsis* and PHŪSIS.

Conclusion

Getting to this point has relied significantly on an analysis of the Aristotelian concept of *dūnamis* and its relation to *phūsis*. But this is not all that needs to be said on *dūnamis* and *phūsis* because their relationship is important also for the discussion in the next chapter where it will be shown that the concept of *conatus* associated with Spinoza derives from an interpretation of *dūnamis*. This is important for environmental philosophy because some significant recent contributions have, in part, relied on a specifically Spinozist use of *conatus*. The demonstration of a close relationship between the concept of *conatus* and the Aristotelian concepts of *telos* and *dūnamis*, provides the groundwork for arguments that will be developed later and which will take up the question of what it is to be morally considerable and what it is for which we should take responsibility.

64 Aristotle, Physics 192b 13-14, transl. R.P Hardie and R.K. Gaye in Jonathan Barnes ed. The Complete Works of Aristotle.

2. Persevering in Being: *Conatus* and *Dūnamis*

Introduction

Within contemporary environmental philosophy there are a number of theories that deal with the concept of self-realization—and accompanying values—through a Spinozist conception of *conatus*, but the interpretations of Spinoza's concept vary. *Conatus* is a concept that attempts to explain how a thing—whether a rock, a tree, a butterfly or a mammal—continues to be what it is despite changes. How and in what sense, for example, can an egg, imago and adult of an individual insect be considered the same entity? The changes a thing undergoes may be triggered by an internal or an external impetus, but the crucial issue is how a thing, 'X', retains its 'X-ness' despite the changes that are triggered. That something can, in Spinoza's terms, 'persevere in its own being' is fundamental to it being able to engage with its world, but understanding this perseverance is also necessary for us to understand what it is that has 'interests' that we need to take into account when working out how we ought to behave toward it.

This chapter shows that Spinoza's concept of *conatus* ultimately owes its major debt to Aristotle's *dūnamis* and can be largely expressed in terms of *dūnamis*. This gives an important foothold for the establishment of an environmental metaphysics based on an Aristotelian biological ontology and lays the ground for an ethics of responsibility that encompasses all living individuals. The chapter is in three parts. The first deals with those aspects of Spinoza's intellectual heritage that influenced him in the formation of his concept of *conatus*. This begins by establishing a connection between Cicero's (Marcus Tullius Ciceronis) use of *conatus* and the Greek expression *hormē*, before briefly examining how *conatus, hormē* and related concepts in Aquinas, Descartes and Hobbes influenced Spinoza. The second part compares Spinoza's and Aristotle's taxonomies of volition and develops a connection between some of Aristotle's concepts and Spinoza's concept of *conatus*. This is a weaker connection than that demonstrated in the third section, but covering it helps show the degree of Spinoza's indebtedness to Aristotle. The third part compares Spinoza's concept of *conatus* and Aristotle's concept of change, particularly that part centring on his concept of *dūnamis*.

Origins of Spinoza's *Conatus*

Aristotle's influence on Spinoza was not direct as there is no evidence that Spinoza could read Greek or even had much acquaintance with Greek philosophical sources—although he was well acquainted with the Roman Stoics.[1] However, it is possible to track a route of influence through Cicero, through medieval and Scholastic philosophy, and through Thomas Hobbes (whose own notion of *conatus* was influenced by his reading of Aristotle's *Rhetoric* and *Physics*).[2] From this it will become clear that despite being indirect, the influence was still considerable and that Spinoza's concept of *conatus* is in most respects the same as that of Aristotle's *dūnamis*. That Spinoza was strongly influenced by Aristotelian thinking is well

1 Yirmiyahu Yovel, Spinoza and Other Heretics Vol. 1, The Marrano of Reason, Princeton: Princeton University Press, 1989, p. 36. See also Paul Oskar Kristeller, 'Stoic and Neoplatonic Sources of Spinoza's Ethics', History of European Ideas Vol. 5, 1984; Levy, Ze'ev *Baruch or Benedict, On some Jewish Aspects of Spinoza's Philosophy*, New York: Peter Lang, 1989.

2 Brian Stoffell, Commonwealth and Civility: a Study of Thomas Hobbes, Diss., University of British Columbia, 1984, p. 79.

documented in Wolfson's[3] meticulous work, which argues that the key elements of Spinoza's intellectual heritage are Aristotle, Moses Maimonides and Descartes. Wolfson does imply that Spinoza may have read Aristotle, but does not say which texts he read or in what language. On the other hand, in his biography of Spinoza, Yovel claims Spinoza did not know any Greek philosophy[4] so it is worth investigating further the nature of any influence.

The concept of *conatus*, which is central to Spinoza's metaphysics, appears most distinctly in the third section of his *Ethics* where he sets up his psychology in terms of three primitive affects (*affectus*): desire, joy and sadness.[5] Within the systematized psychology Spinoza develops in Part III of the *Ethics* strict laws are presented for human behaviour and these are laws that can be organised in such a way that successively 'less general laws...are derived from more fundamental principles'.[6] The chief law in Spinoza's systematization is that 'Each thing, as far as it can by its own power, strives to persevere in its own being'. This is the so-called '*conatus* doctrine' which requires not merely that things strive for self-preservation, but also that they strive to increase their 'power of action'. From this Spinoza deduces other principles. Curley notes that 'the *conatus* doctrine functions not only as a foundation for psychology, but also as a foundation for ethics in so far as it gives content to the notion of a rational plan of life'. Taking Spinoza's definition of the good as 'what we certainly know to be useful to us' we can, in Edwin Curley's words, 'identify what is truly useful to us with what helps us to persevere in our being and increase our power of action, for these are ends we necessarily have'. Our efforts to

3 Harry Austryn Wolfson, The Philosophy of Spinoza: Unfolding the Latent Processes of His Reasoning, New York: Schocken, 1934, 1969. Two volumes.

4 Yovel (Spinoza and Other Heretics).

5 J.A. Simpson and E.S.C. Weiner eds., The Oxford English Dictionary, 2nd ed., Oxford: Clarendon, 1989.

6 Edwin Curley, A Spinoza Reader: The Ethics and Other Works, Princeton: Princeton University Press, 1994, p. xxx.

persevere in our being—our striving—are, however, not conscious, but we can be said to possess 'the good' to the extent that our lives are influenced by 'affects' connected to the striving.

It is commonly held that Spinoza developed his concept of *conatus* from the Scholastic tradition, beginning with St. Thomas Aquinas, a tradition that itself was strongly influenced by interpretations of Aristotle. However, while this is true in part, it is not the whole of the story. To flesh out the rest of the story, it will pay to go back to some of the early uses of the term.

Conatus came into English from Latin about the middle of the seventeenth century. The Oxford English Dictionary says *conatus* has two main senses. Firstly it is 'an effort or endeavour' and was first used in this sense in 1722 by Wollaston in reference to 'natural religion'. The second sense is of 'a force, impulse or tendency simulating a human effort—a nisus'. It was first used in this sense in 1665. Paley's use of the term in his Natural Theology of 1802 is significant because of his importance to Darwin and the positioning of the notion firmly within natural science. The second sense of the term (as used in English) is the one cited in Funk and Wagnall,[7] which defines *conatus* as 'an inherent tendency in animals and plants simulative of efforts in humans'. The inclusion of the idea of an inherent tendency is significant, but it should be made clear that humans have the same sorts of inherent tendencies as other things, although they may also have additional tendencies. There is a human *conatus*, which has importance in Spinoza.

In Latin, there is a long history of the term going back to Cicero. The Oxford Latin Dictionary notes three main senses of *conatus*: a) an effort to accomplish a desired end, endeavour, exertion; b) the same with an emphasis on physical effort, and c) an instinctive effort, an impulse.[8] This third sense seems closest to the Spinozist sense since for Spinoza *conatus* need not involve consciousness. An entity can

7 Isaac K. Funk ed., A Standard Dictionary of the English Language, New York: Funk and Wagnalls, 1901.

8 P.G.H. Glare ed., The Oxford Latin Dictionary, Oxford: Clarendon, 1982.

have *conatus* without being conscious and, in fact, for Spinoza, all objects have *conatus*, although, as will be seen later, there may be different kinds of *conatus*. All the references for the third sense of *conatus* in the Oxford Latin Dictionary are to passages in Cicero. Indeed, it is Cicero's work that is most often cited as a source for examples of the term[9] and for this reason, as well as Cicero's role as a major agent for the transmission of Greek thought into Latin, it is worth examining some of his uses of the term in more detail.

Cicero, Conatus and Hormē

Cicero (Marcus Tullius Ciceronis, 106-43BCE) was an orator, lawyer and politician who also wrote philosophical works. His commentaries give great insight into the Roman empire, but it is his contribution to philosophy that is of interest here. He made important contributions to philosophy in part by translating Greek philosophy into Latin and in the process providing both contemporary and future Latin scholars with some important philosophical vocabulary. He translated works from the then three dominant schools of Greek thinking (Academic, Stoic and Epicurean) and, although he was by training an Academic Sceptic, he also admired the ideas of the Stoics. Cicero made only occasional remarks about Aristotle, whom he interpreted in general as a Platonist. His interpretation of Aristotle in this way was probably due to sympathy for Stoic moral philosophy and to the fact that he was influenced by the dispute between Antiochus and Philo of Larissa at the end of the Academic School. Stoicism is the most likely origin of Cicero's use of the term *conatus*, but in drawing links between Cicero's *conatus* and Aristotle's key concept of *dūnamis*—as we are doing here—it should be borne in mind that the Stoic origins of Cicero's *conatus* imply a different metaphysics, psychology and

9 See also for example: D.P. Simpson ed., Cassell's New Latin-English English-Latin Dictionary, London: Cassell, 1959; William Smith ed. A Smaller Latin-English Dictionary, London: John Murray, 1879.

concept of teleology from those associated with Aristotle. However, it is still possible to point to some relevant links.

Links to Aristotle can be established, for example, in Cicero's De *Natura Deorum* where (at 2.122)[10] *conatus* (as *conatum*) is used to deal with the wonders of animal life, particularly the general adaptation of animal nature for the preservation of the individual. In this passage *conatus* is juxtaposed with both *sensum* (perception) and *appetitum* (appetite) which relates to the distinction between harmful and health-giving, or salutary, things.[11] The connection with *sensum* is discussed below. The relationship between appetite and *conatus* is important in understanding the complexities of *conatus* and will be explored in more detail in the second part of this chapter.

In his translation of *De Natura Deorum*, Pease notes that the cited passage (2.122) refers to Aristotle's *De Sensu* (1, 436a6-11). In the *De Sensu* passage (in which familiarity with the account of the soul given in *De Anima* is assumed) Aristotle writes:

> The most important characteristics of animals, whether common or peculiar, are clearly those which belong to both soul (*psychēs onta*) and body, such as

10 There are a number of critical editions of Cicero's De Natura Deorum (ND). The following editions will be referred to: De Natura Deorum, Libri Tres with commentary by J.B. Mayor, Cambridge: Cambridge University Press, 1883 (Mayor edition); De Natura Deorum, Libri 2 & 3, Arthur Stanley Pease ed., Cambridge, Mass: Harvard University Press, 1958 (Pease edition); Scripta Quae Manserunt Omnia Fasc. 45, De Natura Deorum W. Ax ed., Stuttgart: B.G. Teubner, 1961 (Teubner edition, the publisher being cited because it is part of an edition of the complete works of Cicero).

11 ND 2.122 'Dedit autem eadem natura beluis et sensum et adpetitum, ut altero conatum haberent ad naturales pastus capessendos, altero secernerent pestifera a salutaribus. Tam vero alia animalia gradiendo, alia serpendo ad pastum accedunt, alia volando, alia nando, cibumque partim oris hiatu et dentibus ipsis capessunt, partim unguium tenacitate arripiunt, partim aduncitate rostrorum, alia sugunt, alia carpunt, alia vorant, alia mandunt. Atque etiam aliorum east humilitas, ut cibum terrestrem rostris facile contingant'.

sensation, memory, passion (*thūmos*), desire (*epithūmia*), and appetite generally (*orexis*), and in addition to these pleasure and pain [12]

The three terms *thūmos*, *epithūmia* and *orexis* refer to Aristotle's taxonomy of volition, a taxonomy on which, according to Wolfson, Spinoza's was modelled. The importance of these terms will be explored later in this chapter, but for the moment it is worth noting that there is, *prima facie*, a link between the Greek terms in parentheses, above, and Cicero's use of *conatus*.

A link between Greek terms and concepts and Cicero's use of *conatus* is obvious also at *De Natura Deorum* 2.58, where *conatus* and appetite are both considered species of that which the Greeks called *hormē*—roughly speaking to set in motion, or an impulse. For Cicero *conatus* and appetite were synonyms for *hormē*. That *conatus*, appetite and *hormē* are seen as synonymous is confirmed in Teubner's edition of Cicero's works, which shows that Cicero translated the Greek *hormē* sometimes as *conatus* and other times as *adpetitus* or *adpetitio*, terms that are effectively identical to *appetitus* and *appetitio,* respectively.[13] Cicero's translation practice was not particularly rigorous and he lacked an adequate philosophical vocabulary in Latin so he sometimes created Latin terms and thereby distorted the Greek concepts he was translating. This aside, *De Finibus, De Natura Deorum* and *De Officiis* are generally regarded as an excellent translation of Greek philosophy into Roman terms.[14]

Equivalence between *conatus* and appetite can be seen in numerous places in Spinoza's *Ethics*. For example, he writes (at

12 Aristotle, De Sensu (On Sense and Sensible Objects), transl. W.S. Hett, Cambridge, Mass: Harvard University Press, 1986 (Loeb edition).

13 e.g. at De Natura Deorum 3.33 appetitio (adpetitio) is used for hormē. There is no philological or philosophical difference between appetitio and adpetitio. In a 1933 edition of De Fato, the editor points out that Cicero renders hormē both by the abstract singular adpetitus, by its more concrete plural, and by the concrete adpetitio. Pease edition, p.615n.

14 Jason L. Saunders, Greek and Roman Philosophy after Aristotle, New York: Free Press (Macmillan), 1966.

EIIIP27Cor3D) that 'we strive to free the thing we pity from its suffering' and that 'this will or appetite to do good … is called benevolence' (*E*IIIP27Cor3DS) where the italicised 'or' denotes equivalence.[15] Later in the *Ethics* Spinoza writes: 'striving, or appetite' (*E*IIIP35D); '...by the striving to persevere in one's being, in so far as it is related to mind and body together, we understand appetite and desire'; 'by striving we understand desire' (Ethics IIIP58D) coupled with the equivalence of appetite and desire (at *Ethics* IIIP9S) and; 'by the word desire I understand any of a man's strivings, impulses, appetites and volitions' (*Ethics* III IExp).

The content of the passage at *De Natura Deorum* 2.58 can be traced to Aristotle's *Parts of Animals* (640a22-23) and *Generation of Animals* (734b22-24) where Aristotle refers to the 'creative and formative principle of the seed'. At 734b22-24 this 'creative and formative principle' is, specifically, a '*dūnamei*'. The sense of this is repeated at 734b36 and again at 735a2-4. As will become clear, there are equivalence relationships between *hormē, dūnamis* and the various terms for appetite and desire that indicate a strong provenance in Aristotle for Spinoza's key concepts.[16] Peck translates the passage at 734b22 as: 'Whatever is formed either by Nature or by human Art, say X, is formed by something which is X in actuality (*energeia*) out of something which is X potentially (*dūnamei*).'[17] That is, for Aristotle, animals come to be what they are because of the innate potential (*dūnamis*) contained in the seed. There is clearly a strong connection between *conatus* in Cicero, the Greek term *hormē* and

15 This passage, and those that follow, are from translations by Edwin Curley who notes that the word 'or' translates either sive or seu.

16 Spinoza's and Descartes' accounts of the passions are also indebted to the Neostoicism of the late 16th Century, particularly to that of the Belgian Justus Lipsius (Joost Lips) who published De Constantia in 1584. See e.g. Papy, Jan. 'Justus Lipsius'. The Stanford Encyclopedia of Philosophy (Fall 2004 edition). Edward N. Zalta (ed.), URL = http://plato.stanford.edu/entries/justus-lipsius/#Rel

17 Aristotle, Generation of Animals, trans. A.L. Peck, Cambridge, Mass.: Harvard University Press, 1990 (Loeb edition).

Aristotle's use of *dūnamis*. And, although this brief account makes a too-rapid equation between *dūnamis* and *hormē* it does indicate a connection that is worth exploring further.

A useful place to start is with Wolfson's work on Spinoza because Wolfson[18] argues explicitly that *conatus* can be traced to the Greek term *hormē*. However, as a brief examination of Cicero has shown, *hormē* has also been translated by other terms in addition to *conatus*. There is, apparently, a more complex relationship between *conatus* and its antecedents than one of mere identity. To begin to unravel that relationship, it will pay to examine Wolfson's argument, for although Curley in his early work claims that Wolfson's is a 'fundamentally wrong' way of dealing with Spinoza's formulas[19] there is a vast body of careful scholarship to be drawn on in Wolfson's two-volume work of which even Curley makes use, especially in his later work.

Wolfson begins his coverage of Spinoza's concept of *conatus* and the 'primary passive emotions' by noting that increase and diminution 'imply a certain standard of measurement'.[20] The standard of measurement for Spinoza is the '*conatus* (effort, impulse) by which each thing endeavors to persevere in its own being.' This endeavour for self-preservation is Spinoza's first law of nature. This principle, as the first law of nature, has a long history which is commonly held to have begun with the Stoics, for whom 'an animal's first impulse (*protēn hormēn*) ... is to self-preservation'.[21] According to Cicero however, this view is nothing but a repetition of the Peripatetic view, which in his restatement reads: 'Every natural organism aims at (*vult*)

18 Wolfson, Philosophy of Spinoza, Vol. 2, p. 195.

19 Edwin Curley, 'Spinoza's Metaphysics: an Interpretation', Diss., Duke University, 1963 [UMI 64-4732] ms p. 5.

20 Wolfson, Philosophy of Spinoza, Vol. 2, p. 195.

21 Diogenes Laertius, De Vitis, Dogmatibus et Apophthegmatibus Clarorum Philosophorum Book Z, 85. www.mikrosapoplous.gr/dl/dl07.html #Zinon. Accessed Nov. 2008

being its own preserver.'[22] Cicero, however, is not a neutral witness on this point. As noted earlier, Cicero was a follower of Philo of Larissa and, following Philo, Cicero claims Stoic doctrine as Peripatetic and Academic in origin because the Academy had by this date assimilated much Stoic theory. In keeping with its conceptual origins in Aristotle the 'aim' referred to here is an internal aim.

Wolfson charts a brief history of this internal aim, this notion of 'endeavouring to persist in being', beginning with Augustine who, in *De Civitate Dei* (XI, 27), shows how all things in nature wish to exist or 'conserve their existence'. He then proceeds to St. Thomas Aquinas who reproduces this principle in the statement that 'every natural thing aims at (*appetit*) self-conservation.'[23] Wolfson notes, first that Duns Scotus had a similar position; second, that 'Dante expresses the same sentiment, i.e. 'everything which exists desires (*appetat*) its own existence'';[24] and third, that Telesius held that all things aim (*appetens*) at self-preservation.[25] Wolfson recognizes that a number of Renaissance philosophers also provide statements of the principle of self-preservation and that by Spinoza's time the principle had widespread popular currency, so much so that Spinoza's teacher, Rabbi Saul Levi Morteira began a sermon with 'Nature, mother of all created beings, has implanted in them a will and impulse to strive for their self-preservation.'[26] For all these thinkers, Wolfson observes, 'self-preservation is spoken of as a sort of wish or will or desire expressed by such terms as *vult, velle, appetit*' each of which can be traced to the Greek *hormē*. But he says that '*hormē* can also be

22 De Finibus Bonorum et Malorum, IV, 16. 'Omnis natura vult esse conservatrix sui…', www.thelatinlibrary.com/cicero/fin4.shtml, accessed Nov. 2008.

23 Questiones Disputatae de Potentia, Quaest. V, Art I, 13. Cited by Wolfson.

24 De Monarchia, I, 13 (or 15). The Latin phrase in Dante is Omne quod est appetat suum esse. Cited by Wolfson, p. 196n.

25 De Rerum Naturae, IX,3, beginning.

26 Wolfson, Philosophy of Spinoza, p. 196.

translated by *conatus*' and that Cicero treats *conatus* and *appetitio* as synonymous Latin equivalents of *hormē*.[27]

In all this, however, it should be remembered that this 'will, or wish or desire' is not necessarily psychological desire but is a potential—what Aristotle would call *dūnamis*—the presence of which allows psychological desire, for example, to be manifested. Versions of this potential, with a clear relation to Aristotle, can also be found in St Thomas Aquinas whose influence on Scholastic and early modern thinking is hard to overstate.

St. Thomas Aquinas

The concept of an internal motive force, principle or potential in Aristotle was picked up by St Thomas Aquinas and subsequently passed on through the Scholastic tradition to the early modern period in which Spinoza was writing. This Aristotelian legacy can be found in several places in Thomas, most notably in the *Summa Contra Gentiles*. For example, in the first chapter of the second book of *Summa Contra Gentiles* (para 2) Thomas uses Aristotle's *Metaphysics* (9 viii 9) to help make his argument. In the Dominican translation Thomas writes,

> The operation of a thing is twofold [Arist. Metaph. 9 viii 9; 1050a]...one that abides in the very worker and is a perfection of the worker himself, such as to sense, to understand, and to will; and another that passes into an outward thing, and is a perfection of the thing made that results from it, such as to heat, to cut, and to build.[28]

James F. Anderson translates this as,

27 Ibid.

28 St. Thomas Aquinas, The Summa Contra Gentiles of St. Thomas Aquinas, Books I and II, Literally translated by the English Dominican Fathers from the latest Leonine edition, London: Burns, Oates and Washbourne, 1924.

> There are, however, two sorts of operation, as Aristotle teaches in *Metaphysics* IX (1050a25): one that remains in the agent and is a perfection of it, as the act of sensing, understanding, and willing; another that passes over into an external thing, and is a perfection of the thing made as a result of that operation, the acts of heating, cutting and building for example.[29]

In his introduction, Anderson writes, with reference to this passage:

> ...the former type of action is called 'immanent', and thus belongs to the consideration of God in Himself; while the latter is called transient, and therefore pertains to the problem of creation...the study of God...in his works.

But Thomas' God is not Aristotle's God. The whole activity of Aristotle's God is self-contemplation and while being perfect activity/actuality he is independent of nature, but, because he is the *telos* that nature imitates, nature depends on him.

In Aristotle's terms a potential (*dūnamis*) to act is present in the agent. This potential to act can be mobilised to act on the agent itself, qua itself, or it can be mobilized to act on and thus change in some way something external to the agent. This potential has an active and a passive form. The first sort of action referred to in the above passage is internally generated and is the agent acting on itself qua itself. This is possible because an active *dūnamis* within the agent interacts with a like passive *dūnamis* also within the agent. The second aspect of action can be described in terms of an active *dūnamis* in the agent acting on an unlike passive *dūnamis* in the 'patient', an object external to the agent.

An appeal to the Aristotelian notions of potential, movement and actualization occurs also at *Summa Contra Gentiles* I, xvi where Thomas says,

> actuality is simply before potentiality: because potentiality does not bring itself into actuality...Just as it is natural that a thing should act in so far as it is actual,

29 St. Thomas Aquinas, Summa Contra Gentiles, transl. James F. Anderson, Book 2: Creation, Notre Dame: University of Notre Dame P, 1975 - first pub 1956.

> so it is natural for it to be passive in so far as it is in potentiality, *for movement is the act of that which is in potentiality.*

The claim that actuality is simply before potentiality can be seen in Aristotle's *Metaphysics* (1050b4) but the passage cited by Thomas and indicated by italics in the preceding quote is probably that at *Physics* 201b 5-6 which Wicksteed and Cornford translate as 'it is clear that motion (*kinēsis*) must be the realization (*entelecheia*) of the specific potentiality (*dunatoū*) in question and of the subject only qua seat of this specific potentiality (*dunatōn*).'[30] Thomas is using Aristotle to demonstrate, in part, the necessity of a Prime Mover. In the Aristotelian scheme, actuality is ultimately (*eschaton*) before potentiality in that, in a regression of actions and potentials, the end point is ultimately a Prime Mover (in Thomas' case the Christian God) which is pure actuality/activity—a claim made clearly at *Metaphysics* 1050b5-6 where Aristotle states 'one actuality presupposes another right back to that of the prime mover.' Aristotle's articulation of the necessity of a prime mover is well understood but there is, however, a controversy over Aristotle's definition of motion here. This controversy is set out by Gill, who asks how the definition [of change] can be interpreted 'so that the actuality is the process rather than the product'[31] and by Kosman who asks, 'When Aristotle says that motion is the actualization (*entelecheia*) of the potential...does he mean a process or a product?[32] This is not an easy question to address. W.D. Ross would have it that Aristotle, especially at *De Anima* II.5, distinguishes between two sorts of potentiality—one for being and one for becoming. Gill argues against this on the basis that a potentiality for becoming is a metaphysically derivative category because it admits reduction to more basic components. Kosman also claims Aristotle is concerned only with potentiality for

30 Aristotle, Physics, transl. Philip Wicksteed and Francis Cornford, London: Wm. Heinemann, 1929 (Loeb edition).

31 Gill, Aristotle on Substance, p. 185.

32 L.A. Kosman, Aristotle's Definition of Motion, Phronesis, Vol. 14, 1969, p. 40.

being and not with the potentiality for becoming[33] and subsequently claims that Ross's view on this point is substantially that of Thomas in the *Commentaria*, especially numbers 3 and 5. For Gill Aristotle's definition of motion (e.g. at *Physics* 201b4-10) specifies change as something dynamic and directed and that for Aristotle, 'change is...the actuality of the subject once it has become active and only while it remains active'. But, she asks, How can Aristotle capture this dynamic quality?[34] The answer lies in his two terms *energeia* and *entelecheia* which are sometimes treated as near-identical synonyms, but the etymology of which makes clear they are different.[35] To avoid circularity in his version of change in Aristotle, Kosman highlights a distinction between *energeiai* (activities) and *kinēseis* (changes) and so also captures a dynamic (process) quality in Aristotle's idea of change. Kosman's re-definition has change as the actuality (*entelecheia*) that is the activity (*energeia*) of something that is potentially other than it actually is. Gill also says (referring to Kosman) that changes differ from activities, especially in relation to ends, in the following way because change is concerned with ends without itself being an end. When change reaches an end then the change is completed so change stops. Activities, however, are ends in themselves and don't aim at goals beyond themselves. An 'activity is complete at any moment for as long as it lasts.'[36]

Kosman's thesis enables Aristotle's change to be characterized as dynamic, but for Gill the thesis is inadequate 'because change [in Kosman's proposal] is not only dynamic but also directed toward a goal that the subject fails to possess.' Her solution is to see change as 'the mutual actuality of an agent and a patient, which is located in the patient.' Change is located in that which is moved and exists only so long as the agent affects the patient and the patient responds.[37] Of

33 Ibid. p.44.

34 Gill, Aristotle on Substance, p.192.

35 For a strong discussion of this, see George Blair Energeia and Entelecheia.

36 Gill, Aristotle on Substance, pp. 192-193.

37 Ibid., p. 194.

course, the agent here could be internal or external: it could act on itself *qua* itself, on itself *qua* other or act on another external to itself. An agent acting on itself *qua* itself would in effect have its end within it, a position in accord with Blair's translation of *entelecheia* as 'having the end within'[38] and also consistent with one sense of Spinoza's *conatus*. An agent's ability to act on itself *qua* itself is an element that Aristotle's teleology and Spinoza's *conatus* have in common and that is, moreover, a defining characteristic of a living organism.

The expression in Thomas' *Summa Contra Gentiles* II translated as 'abides in' (in the Dominican translation above) also appears to be picking up on an active sense of Aristotle's terms *energeia* and *entelecheia*—a sense that does not come out in the Oxford translations, but becomes clearer in Blair's—a sense of internal or immanent activity that could be understood as a process rather than a static quality of an entity. Thomas notes that the aspect of the 'operation' of a thing that 'abides in the worker' also 'claims for itself the name...of action.'[39] For Thomas, just as for Aristotle, there is an internal or immanent action or operation which makes an external action possible.

In the excerpt from *Metaphysics* (1050a 12-14) cited earlier Thomas picks up on Aristotle's building example[40] which Tredennick translates as '...men possess the art of building in order that they may build'.[41] From this it could be said that there is something in the person that allows (and so must precede) an external activity. That something in the person is a *dūnamei*, the moving cause of the activity of building. This passage in the *Metaphysics* is preceded immediately by one which makes the point in more general terms. The Tredennick

38 George Blair, Energeia and Entelecheia, p. 81.

39 Thomas, Summa Contra Gentiles II,1.

40 Aristotle, Metaphysics, transl. Hugh Tredennick, London: Wm. Heinemann, 1936; Cambridge, Mass.: Harvard University Press (Loeb edition) 1050a 12-14.

41 Ibid., 1050a12.

translation (which follows substantially the work of W.D. Ross) has this prior passage as

> ...everything which is generated moves towards a principle, i.e. its end. For the object of a thing is its principle; and generation has as its object the end. And the actuality is the end, and it is for the sake of this that the potentiality is acquired.[42]

The translation does not follow the Greek literally but the italicized terms refer to the Greek terms *energeia, hou heneka* and *dūnamis* respectively. The expression *hou heneka* (discussed in the previous chapter) occurs twice in the Greek but in Tredennick's translation the first reference is deleted and the second reference refers not to the 'for the sake of which' but says that it is for the sake of the actuality-as-the-end that potentiality is acquired. Tredennick's translation of *energeia* as actuality highlights the confusion over understanding of *energeia* and *entelecheia.*

There is, evidence[43] that the terms *energeia* and *entelecheia,* which Aristotle coined and used widely, have been mistranslated in the standard translations. This mistranslation serves to attach a static character to the terms instead of the dynamic character that is more in keeping with the term *dūnamis* , to which *energeia* and *entelecheia* are closely correlated. Blair's argument that the primary meaning of *energeia* involves *kinēsis*, i.e. that *energeia* has a more dynamic character, is borne out by a passage at *Metaphysics* 1047a33-4 that Tredennick translates as 'for it is agreed that actuality (*energeia*) is properly motion (*kinēsis*)'. However, moving away from the static translation and choosing to translate *energeia* as internal activity is not unproblematic and there is even a danger that it may actually be misleading. There is dispute, for instance, over the degree to which *energeia* and *kinēsis* are related. It should be made clear then, that when he uses *energeia*, Aristotle isn't talking about activity internal to

42 Ibid., 1050a8-11.

43 e.g., Blair who provides a full catalogue of where the terms are used in Aristotle.

the body or even the embodied self, he is talking about an activity that neither ends nor has a detachable product. He is talking about a process that is integral to and—possibly—definitive of a living thing.

Energeia was the first of the twin terms to be coined by Aristotle. It first appeared in the early work *Protrepticus* (written while Aristotle was still at the Academy with Plato)[44] where it was applied only to life and the living. [45] Its application only to the living here is an important point for, although Aristotle later applied the two terms widely, taking in metaphysics, physics, psychology, ethics and rhetoric as well as biology, the original biological application of the concepts is never repudiated (see e.g. *Metaphysics* Θ).

Blair translates *energeia* as: 'to be internally active',[46] 'immanent activity'[47] or 'internal activity',[48] arguing that the sense of internal activity is borne out by the etymology of the term as Aristotle's neologism *energeia* was made up from the prefix en- meaning 'in' or 'within', and *ergein* 'a rare active voice of the common verb *ergazesthai* which as a middle deponent has the active meaning of 'to work,' 'to do,' 'to act,' or 'to be busy'.'[49] The active voice form *energein* and the noun *energeia* derived from it both originated with Aristotle. Blair argues that Aristotle's choice of a rare active voice for the root of his new word was to 'underscore the verb's active force' and that 'from the very derivation of the word, it would seem that Aristotle was going out of his way to avoid having his word interpreted as 'actuality' or as a kind of static modality of being'.[50] An English language analogy, Blair says, might be drawn from the rare active voice expression 'to busy oneself about' so that, to a Greek, Aristotle's noun would correspond to the unusual word 'busyingness',

44 Blair, Energeia and Entelecheia, p.20, citing Jaeger and Ross
45 Ibid., p. 25.
46 Ibid., p. 27.
47 Ibid., p. 29.
48 Ibid., p. 53.
49 Ibid., p. 17.
50 Ibid., p. 18.

which would not be interpreted in a static sense.[51] Adding the *en-* to *ergeia* would create a word meaning etymologically 'to be busy within' or 'to be internally active''. This new noun (*energeia*) is close in meaning to the phrase 'immanent activity', that was used by the Scholastics to indicate a necessary characteristic of living beings[52]—beings that have 'action that begins and ends in the agent'[53] as opposed to action that starts in an agent and 'ends in a patient who is not also that agent.[54] Blair notes also that 'Aristotle is looking for a 'working' that doesn't work on anything, but is just being active inside the agent.' These same distinctions are made by Gill,[55] citing especially Aristotle *Metaphysics* Θ.1 and Δ.12.

There is in *energeia* a sense of reflexive activity—where an agent acts on itself. This makes it one of the defining characteristics of a living entity in which the entity acts on itself *qua* itself in a process of self-maintenance and self-production. Aristotle's key terms had their origins in his biology and apply primarily to living things, even though they also have non-biological applications.

Blair says *entelecheia* was first used later than *energeia*, but despite this, Aristotle stopped using it because its meaning could be subsumed by that of *energeia*. He says Aristotle invented *entelecheia* to address the ability to be instead of to act[56] and that although Aristotle had used it extensively he eventually realized that ' *entelecheia* was a superfluous term'[57] because in all cases existing was in fact an internal activity.

For Aristotle, who was intensely interested in things biological, living things had something special that made them 'living' and the

51 Ibid., p. 19.

52 Robert Brennan, Thomistic Psychology: a Philosophic Analysis of the Nature of Man, New York: Macmillan, 1941, pp. 106-8.

53 Blair, Energeia and Entelecheia, p. 19

54 Ibid, p.19.

55 Gill, Aristotle on Substance, p. 173.

56 Blair, Energeia and Entelecheia, p. 77.

57 Ibid., p.154.

etymology of *entelecheia* gives a clue to what he might have thought that special thing was. The etymology, as Blair sees it, indicates very strongly that Aristotle saw an internal *telos* as an essential part of any living entity. He argues that the *en-* means 'within' or 'internal', the *tel-* comes from *telos* meaning 'end' and the final part of the word comes from *echon* 'having' (the present participle is preferred because the Greek present tense is closer to the English present continuing tense). A translation based on the etymology used by Blair would produce a word meaning 'having the end within [the being]'.[58] It is consistent with what is known about Aristotle, particularly his interest in biology and the background in medicine he gained from his father Nicomachus, that Aristotle would have noticed that when living things act, they aren't necessarily responding to some action on them, and nor are they always doing something *to* something either. That is, he noticed that there is some *internal* motive force unique to living things. It is this concept of an internal motive force in Aristotle that was picked up by Thomas and subsequently passed on through the Scholastic tradition to the early modern period in which Spinoza was writing. The links between Aristotle and Thomas are well documented, but the reliance on 'static' translations of Aristotle's key terms (into English at any rate) may have diminished an appreciation of the debt that Thomas's sense of *internal or immanent activity* actually owes to Aristotle. This in turn may have obscured links between internal activity in Aristotle and other expressions of an internal motive force, such as Spinoza's notion of *conatus*.

Descartes

Spinoza was much influenced by Descartes. Some, such as Gilles Deleuze,[59] contend that the influence was largely negative because

58 Ibid., p. 79.

59 Gilles Deleuze, Expressionism in Philosophy: Spinoza, transl. Martin Joughin, New York: Zone Books, 1992.

Spinoza developed a philosophy that was significantly anti-Cartesian. Alan Donagan,[60] however, argues that Spinoza's philosophy was fundamentally dualist and so owed a positive debt to Descartes. It is certain, though, that Spinoza did take much from Descartes including, for example, the principle that there are only substances and modes.[61] There is also some evidence to suggest Spinoza was influenced by Descartes' idea of tendency in the development of his *conatus*, although there is dispute about the strength of the evidence. On the supposed links with tendency, David Lachterman, for example, says the notion of *conatus* (although not the term) can be found in Descartes' idea of a body's tendency to move in a straight line, which Spinoza deals with in the *Cogitata Metaphysica*. Further support for the view that Spinoza drew on a Cartesian concept of *tendency* can be gathered by coupling the fact that Hobbes' notion of *conatus* owed a debt to Descartes' *tendency*[62] with the fact that (as will be seen below) Spinoza was influenced by Hobbes' notion of *conatus* in the formation of his own. This idea of the tendency to move in a straight line, Lachterman says, was viewed in Descartes' time as being suspiciously like 'a reversion to the Scholastic notion of potentiality or virtuality'.[63] As shown earlier, this notion of potentiality originated with Aristotle's *dūnamis*.

The fact that the Scholastic notion of potentiality is derived largely from Aristotle's notion of *dūnamis* provides *prima facie* evidence of genealogical link between Aristotle and Spinoza on this point. The link is indirect, but still adds weight to the claim that

60 Alan Donagan, Spinoza's Dualism, Richard Kennington ed., The Philosophy of Baruch Spinoza, Washington D.C.: Catholic University of America Press, 1980, pp. 89-102.

61 Spinoza, Cogitata Metaphysica II.V, trans. Edwin Curley, Complete Works of Spinoza.

62 McNeilly, Egoism in Hobbes, Philosophical Quarterly, Vol. 16, No. 64, 1966, pp. 193-206. Cited by Stoffell, 'Commonwealth and Civility'.

63 David R. Lachterman, The Physics of Spinoza's ETHICS, Robert J. Shahan and J.I.Biro eds., Spinoza: New Perspectives. Norman: University of Oklahoma, 1978, pp. 71-112.

Spinoza's *conatus* owes a significant debt to Aristotle. Lachterman claims, without argument, that the suggestion that Descartes is reviving a Scholastic notion 'is of no help because *conatus* is derived from the 'abstract' notion of inertia'.[64] He suggests that Descartes' kinematic theory is at odds with his dynamics and that Spinoza 'eliminates this incompatibility by adopting a strategy yielding a single ground or principle both for genuine physical change and movement, namely, the principle of *conatus*'.[65] Lachterman does not, however, mention that Spinoza's physics has clear similarities with Aristotle's on this point—something taken up later in this chapter. But some clarification of Lachterman's suggestion can be found in Curley, who says that in Spinozistic-Cartesian physics the expression '*conatus ad motum*' is generally held to refer to the tendency of things to 'obey' the principle of inertia 'with no implication of there being any psychic state present'. Curley maintains, however, that it would be incorrect to translate *conatus* as 'tendency' because Spinoza had made a deliberate decision to adopt a term that suggests a psychic state is present. Curley goes on to say in the same passage that 'the *conatus* by which each thing 'strives' to persevere in its being is best regarded as a metaphysical generalization of [the] physical principle of inertia.'[66]

But as will be made clear later in this chapter, Spinoza's *conatus* has both a psychic and a non-psychic dimension.[67] That is, an entity need not be alive to have the *conatus* to persist in being—but those that are alive have an additional form of *conatus*. So, all entities have a *conatus* simply from being extended substance—this might be called

64 Ibid., p. 81.

65 Ibid., p. 82.

66 Curley, Complete Works of Spinoza, p. 657.

67 See e.g., Spinoza Ethics IIIP28D '...the mind's striving, or power of thinking, is equal to and at one in nature with the body's striving or power of acting...Therefore we strive absolutely, or want and intend that it should exist.' The two forms of striving are clear, although the 'psychic' conatus is (only) a manifestation of natural (absolute) striving.

‘bodily’ *conatus*—whereas those entities that are alive have, additionally, what might be called a ‘psychic’ *conatus*, although such *conatus* does not require consciousness. A rock has *conatus*, it strives to persist in its own being, but a living thing has an additional *conatus* to persist in its being.

Because of the growing dominance of Cartesian thought in Spinoza’s day, Spinoza could hardly fail to be influenced by Descartes, but despite the similarities between Descartes’ tendency and Spinoza’s *conatus*, it is in the philosophy of Thomas Hobbes that Spinoza found the greatest contemporary inspiration for his concept of *conatus*.

Hobbes

Spinoza was no doubt aware of the concept of *conatus* before he read Hobbes, but it is also certain that Hobbes was a significant contemporary influence in the formation of his *conatus* doctrine and that Spinoza was introduced to key Aristotelian notions through the work of Hobbes.

In the Latin version of *Leviathan* published in Holland in 1668, 17 years after the English original, Thomas Hobbes uses the terms *appetitus* and *cupido* as the Latin equivalents of the Greek *hormē* and notes that this is a form of *conatus*.[68] In the English version the equivalence between *conatus* and appetite and desire is represented in the following way:

> the small beginnings of Motion, within the body of Man....are commonly called Endeavour. This Endeavour, when it is toward something which causes it, is called Appetite, or Desire; the later being the generall name.’[69]

68 Wolfson, Philosophy of Spinoza, Vol.2, p. 197.

69 Thomas Hobbes, Leviathan, London: Penguin, 1985 (first pub. 1651), I.6.

For Hobbes, Appetite is the equivalent of *hormē* and is a sub-species of desire, a position that will be seen below to be reflected in Aristotle and in Spinoza.

The publication of *Leviathan* in Latin, in Spinoza's home town in Holland, and in the year 1668,[70] is noteworthy in that it gave Spinoza access to Hobbes' ideas at an important time. At the same time that the Latin *Leviathan* was published in Holland Spinoza stopped work on his *Ethics*, and subsequently produced his *Theological-Political Treatise* which he intended to pave the way, as a popular precursor, to the publication of the *Ethics*. Publication of the *Theological-Political Treatise* did not, however, have the desired effect and Spinoza had to suspend publication of the *Ethics*, which was eventually published only after his death. Spinoza studied Hobbes closely while writing the *Treatise* and the latter part of the *Ethics*, which was completed after the *Treatise*, 'shows quite strongly...the influence of the philosophy of Hobbes'.[71]

Spinoza's use of the concept of *conatus* begins in the third section of the five that comprise the Ethics and continues throughout the rest of the work. Hobbes' *conatus* was not only an element of his political philosophy, but was 'a general theory of dispositions' and owed much to Aristotle.[72] With *The Elements of Law*, for example, Hobbes hoped to 'speak sense' about the 'facultyes and passions of the soul'. What resulted was heavily dependent on Hobbes' reading of Aristotle's *Rhetoric* and *Physics* from which he postulated a form of motion that he called Endeavour in English and *conatus* in Latin to explain how 'the passions could exist as dispositions'.[73] That postulated form of motion (Endeavour or *conatus*) was drawn directly from Aristotle.[74] For Hobbes the primary function of endeavour is 'to explain the

70 C.B. Macpherson, 'Introduction', Thomas Hobbes, Leviathan, London: Penguin, 1985.

71 Curley , Spinoza Reader, p. xx

72 Stoffell, 'Commonwealth and Civility', p. 8.

73 Ibid., p. 79.

74 Ibid., p. 81.

internal beginnings of animal motion'[75]—a function that the concept of *dūnamis* performs in Aristotle.

As mentioned earlier, there is also a much more subtle and indirect Aristotelian influence on Hobbes', and hence on Spinoza's, concept of *conatus*—an influence through Descartes' notion of 'tendency'. The concept of 'tendency' was widely held by Descartes' contemporaries to be a reversion to the scholastic notion of potentiality, which, as we have seen, was itself derived from Aristotle's *dūnamis* .

There are clearly a number of indirect Aristotelian influences on the formation of Spinoza's concept of *conatus* (Scholastic notions of potential, Descartes' *tendency*, Hobbes' *conatus*) in addition to the Aristotelian notions that in all likelihood came to Spinoza in the works of Averroes (Ibn Rushd), Avicenna (Ibn Sinna), Maimonides and now-lost Hebrew manuscripts.

The foregoing by no means demonstrates exactly where and how and under what influences Spinoza came to his notion of *conatus*, but it does establish that there is a historical connection between *conatus* and *hormē* , particularly as employed by Cicero and Hobbes and that these terms can in most respects be considered identical. There is also a strong suggestion that the concept behind both terms had its origins in Aristotle's *dūnamis* and that there is also a likely connection with Aristotle's theory of volition, a connection that is worth examining.

Desire and Will: a Taxonomical Comparison

The notion of desire and the role of the will are important to any understanding of Spinoza's *conatus*, especially that *conatus* which inheres in animate entities. And, as will be seen, will and desire form

75 Ibid., p. 82.

part of the link between Spinoza's concept of *conatus* and Aristotle's concept of internally-generated movement or change.

As shown earlier, *conatus* can be traced to the Greek term *hormē* and particularly to Stoic uses. However Spinoza also adopted something very like Aristotle's taxonomy of volition and it can be shown that his concept of *conatus* owes something to the Greek view of desire, particularly the notion of desire represented by the term *orexis* and its cognates. The link between Aristotle's taxonomy of volition and Spinoza's *conatus* is not as strong as the link between *conatus* and *hormē* and thence to *dūnamis*, but it is still worth investigating in order to get a fuller picture of the connections between Spinoza and Aristotle with respect to *conatus*.

When Spinoza uses the Latin *cupiditas* to refer to one of the emotions or 'affects' it is usually translated as desire. Spinoza's concept of *conatus* is linked to desire (*cupiditas*) and appetite (*appetitus*), a link that forms part of a taxonomy of 'will' or 'volition' that in all probability Spinoza borrowed from Aristotle, whether directly or not[76]. Part of the relationship between *conatus* and volition in Spinoza can be seen in Hampshire's analysis, where he says

> human beings can be said to be aware of the tendency to self-preservation and to the increase of their own power and activity which constitutes their 'actual essence' as individuals; the reflexion in idea of this *conatus*, which is common to all particular things in Nature, is called desire (*cupiditas*). Desire is defined as appetite (*appetitus*) together with consciousness of it; appetite is the *conatus*, or tendency to self-preservation, expressed neutrally, that is, neither in purely mental nor in purely physical terms.[77]

Hampshire adds that Spinoza's theory of *conatus* is a theory of desire and will through which it can be explained that humans pursue pleasure because it is a law of nature that they should do so. Humans necessarily pursue pleasure in the sense that their pleasures and their choices arise from the *conatus* of the human as a biological

76 Wolfson, Philosophy of Spinoza, Vol. 2, p. 165.

77 Stuart Hampshire, Spinoza, London: Faber and Faber, 1956, p. 96.

organism—that is, in terms of its own self-maintenance and self-preservation. [78] Humans are, in other words, a part of nature and the choices humans make arise from the *conatus* of the human organism, the *conatus* that is the tendency to self-maintenance and self-preservation.

In his *Short Treatise*[79] Spinoza reproduces a taxonomy of will or volition that has desire as a genus with two species—the species of will toward good and the species of will toward bad. Curley suggests in a footnote to his translation of the passage that the notion of desire being a genus with two species derives from Aristotle's *De Anima* III, 10, although through Scholastic intermediaries, rather than directly from Aristotle.[80] That this chapter in Spinoza is primarily an exposition of Aristotle's doctrines is supported by Mignini, editor of a critical edition of the *Short Treatise.*[81] However, although Spinoza's account of will may have been strongly influenced by Aristotle's, his account is not a mere copy.

Aristotle's taxonomy of volition or desire in *De Anima* III, 10 has *orexis* as the genus and *boulēsis, epithūmia* and *thūmos* as species. In a note to his translation of *De Anima* (at 414b2), Hicks notes that:

> Though not always consistent, Aristotle in the main adheres to the nomenclature...according to which *orexis* is a generic term for that which appears in the rational soul as *boulēsis* (wish or conation) and in the irrational soul as *thūmos* and *epithūmia.* [82]

At *De Anima* 411a28, Hicks notes that from 433a22 and following passages, 'that *boulēsis* is a species of *orexis* and is opposed to *epithūmia* as rational to irrational desire.' The term *orexis* literally

78 Ibid., p. 101.
79 Short Treatise Section II, Ch. 17
80 Curley, Collected Works of Spinoza, p. 126. The key passage is at De Anima 432b 3-6.
81 Ibid., p. 126n.
82 Aristotle, De Anima, transl. R.D. Hicks, Amsterdam: Adolf M. Hakkert, 1965, p. 332.

means 'striving'. Wolfson says that in Aristotle it has the sense of propension or inclinations. Aristotle, along with Plato, had three species in the genus 'desire' (*orexis*). These three species, as the Hicks passage above indicates, were 'wish' (*boulēsis*), 'spirit' (*thūmos*, concerned with social prestige, pride, insult and vengeance) and 'sensual appetite' (*epithūmia*).

In Spinoza's taxonomy the genus is *cupiditas* (desire) and the species are *voluntas* (the will toward doing good—for Spinoza nothing is good or bad in itself) and *voluptas* (the will or inclination toward something that is bad, non-rational or irrational[83]), the two forms of irrational desire in Aristotle (*thūmos* and *epithūmia*) being compressed into one in Spinoza. It is significant that Spinoza's major division is between rational and irrational, as it is in Aristotle. 'Desire' translates both *cupiditas* and *orexis*, and *voluntas* is a direct translation of *boulēsis*. The meaning of *voluptas* covers both sensual and spiritual satisfaction, enjoyment, pleasure and delight and so might be considered a compression of *thūmos* and *epithūmia*. *Voluntas* is governed by the rational intellect and *voluptas* by the irrational side of the intellect. In short, Spinoza's taxonomy of volition looks to be modelled on Aristotle's in that both have a genus of desire with rational and non-rational species. It cannot be said, however, that Spinoza derived his theory of will directly from the Greek of Aristotle although it is safe to say that Spinoza's taxonomy of desire draws on the strong Aristotelian influence passed down to him through, for instance, Augustine, Thomas Aquinas, Descartes, Hobbes and the Moorish and Jewish intellectual traditions.

From the Aristotelian influence comes a clear distinction between rational and non-rational desire. Aristotle follows Plato in that all desire aims at the particular kind of good which defines a given species of desire. For rational desire the good it aims at is the true

83 It need not be inclination to bad, but may also simply be non-rational or irrational, in two senses: i) for those beings which can feel but not reason and; ii) for reasoning beings who decide to do something bad, particularly to themselves.

good, which can be distinguished by reason. In other cases the good aimed at is the apparent good, which is not necessarily bad—only in excess or in inappropriate circumstances. Spinoza appears to misapprehend this distinction at one place in the *Cogitata Metaphysica* where he opposes what he sees as an Aristotelian position on the difference between will and wanting. This position he states as:

> If the will can will contrary to the last dictate of the intellect, if it can want what is contrary to the good prescribed by the last dictate of the intellect, it will be able to want evil as evil. But the latter is absurd. Therefore the former is also.

He comments on this as follows:

> From this argument it is plain that they do not understand what the will is. For they confuse it with the appetite which the soul has after it has affirmed or denied something. This they have learned from their Master, who has defined the will as an appetite for what seems good.[84]

The Master is the common term for Aristotle in both Scholasticism and the early modern period. Curley[85] suggests that Spinoza might be referring to a passage in the *Rhetoric* (1369a1-4) or be relying on a Scholastic source such as Thomas. It is clear, though, from *De Anima* 433a23-30 that what Spinoza is opposing is only a partial reading of Aristotle's position on *orexis* and the good (*to agathon*). In *De Anima* Aristotle distinguishes between the good, the apparent good and the practical good. Hicks translates the passage as:

> ...rational wish (*boulēsis*) is appetency (*orexis*); and, when anyone is moved in accordance with reason, he is also moved according to rational wish. But appetency may move a man in opposition to reason, for concupiscence (*epithūmia*) is a species of appetency. While, however, intellect (*nous*) is always

84 Spinoza, Cogitata Metaphysica, (Gebhardt I/278) transl. Curley, Collected Works of Spinoza, p. 344. The term 'Master' in this passage refers to Aristotle.

85 Curley, Collected Works of Spinoza, p. 344n.

> right, appetency and imagination (*phantasia*) may be right or wrong. Hence it is invariably the object of appetency which causes motion, but this object may be either the good (*to agathon*) or the apparent good (*to phainomenon agathon*). Not all good, however, but practical good (*to prakton agathon*): where by practical good we mean something which may not be good under all circumstances.

Here Aristotle sets up within desire (*orexis*) a division between rational and non-rational of the sort already noted above, between what is right and wrong, and between what is good and what seems good. Spinoza appears to focus on only one side of the division.

Despite the links that have been established between Spinoza's *conatus*, the term *hormē* and Aristotle's theory of volition, there is no etymological link between *hormē* and *orexis*. Chantraine,[86] for example, notes that *orexis* carries the meanings of 'désir, appétence' in Democritus and Aristotle and that the term is defined at *De Anima* 414b. This passage in *De Anima* deals with the entelechy (*entelecheia*) or internal directiveness of the soul and the powers (*dūnamei*) of the soul—namely those of nutrition, appetency, sensation, locomotion and understanding.

It is clear that there is a link between Spinoza's *conatus* and Aristotle's theory of the will, a link summarized in the lower part of Figure 1, at the end of this chapter. The figure summarizes the interrelations between the key concepts and suggests how Spinoza might have developed his notion of *conatus* on the basis of these influences. It should be emphasised, however, that the major antecedent in Aristotle of Spinoza's *conatus* is the concept of *dūnamis* examined more closely below.

Evidence of the links between Spinoza and Aristotle can only be claimed with certainty to come from secondary sources such as Scholastic texts, medieval Hebrew manuscripts, Averroes, and some of Spinoza's contemporaries and near contemporaries. Curley believes

86 Pierre Chantraine, Dictionnaire Etymologique de la Langue Grècque: Histoire des Mots, Paris: Editions Klincksieck, 1968, Vol. 3, p.187.

Spinoza came to Aristotle's thought through the mediation of the Scholastic philosophers, but Wolfson appears certain that Spinoza read Aristotle himself, as indicated by the passage 'In the writings of Aristotle and in the works which transmitted the teachings of the Stoics he found the origin of those views which are only faintly reflected in the works of the rationalist Jewish and Christian theologians'[87]The works transmitting Stoic thought are presumably those of Cicero and Diogenes Laertius.

In the first volume of his work on Spinoza, Wolfson says that the influence of Aristotle was very strong in the Jewish philosophical tradition—in which Spinoza was well schooled. Spinoza did not learn Latin until an adult but had excellent Hebrew which makes it quite certain that Spinoza's primary knowledge of philosophy was from Hebrew literature and that in learning Latin he simply gained 'a new vocabulary for old ideas'[88]. Jewish philosophy was to be expelled from European thinking, but in Spinoza's time Jewish philosophy had not yet been eliminated and even though Jewish texts were generally available only in manuscript form rather than as books, they were readily available to Jewish scholars and apparently contained much commentary on Aristotle.

Conatus and 'Change' in Aristotle

In the first references to *conatus* in the *Ethics* Spinoza indicates that a thing's *conatus*, or striving to persevere in its being, is something done from the thing's own power (*Ethics* IIIP6) and is nothing but the actual essence of the thing (*Ethics* IIIP7). The power referred to here is a translation of the Latin *potentia* which itself is the standard

87 Wolfson, Philosophy of Spinoza, vol. 2, p. 222

88 Ibid., Vol 1, pp. 12-13.

translation of Aristotle's term *dūnamis* . This idea that the essence of a thing is the internal power by which it persists in being appears, on the face of it, to be an Aristotelian position. However, this *prima facie* appearance needs careful scrutiny. This section aims to determine whether this notion does have an Aristotelian basis by comparing Spinoza's *conatus* and Aristotle's concept of internal movement, centring on Aristotle's concept of *dūnamis* . It begins by looking at Aristotle's general position on change, then looks more closely at the concept of *conatus* and its relation to Spinoza's concept of Nature. The section, and the chapter as a whole, will conclude that Spinoza's *conatus* is strongly indebted to, and is in most respects identical with, Aristotle's concept of *dūnamis*. The relationship between *conatus* and *dūnamis* has important ramifications for environmental philosophy that will be explored in the following chapters.

Four Kinds of Change

For Aristotle, change is an ontologically distinctive feature of natural things. But change in living things has unique features that deserve special scrutiny

In Aristotle's physics there is a single ground or principle for change or movement. The single ground for change is *dūnamis* . As discussed in the previous chapter, for there to be change there has to be both a passive and an active *dūnamis*. That is, for change to occur there must be an active or actualized potential for change acting on a passive potential for change in which like can act on like or unlike.

Aristotle recognized four kinds of change or motion: coming into being and passing away (*genesis* and *phthora*); physical increase or decrease (*auxēsis* and *phthisis*); the change of one substance into another (*alloiōsis*—transformation) and change from one place to another (*phora*). Aristotle makes a concise statement of this at *De Anima* 406a12-13 where he says: 'Of motion there are four species, change of place or locomotion, change of quality or alteration, diminution and augmentation'. These can be summarized as *substantial change, quantitative* change, *qualitative* change and change of *place* (*topos*).[89] The last three are covered by Aristotle's word for motion in the narrow sense (*kinēsis*) whereas all four are covered by the term for his broader concept of change (*metabolē*).

So, at *Physics* 192b19, for example (where *hormēn echei metabolēs* is translated by Wicksteed as 'inherent trend toward change') the use of *metabolē* is significant because in this passage *hormē(n)* is related to the broader concept of change rather than the narrower one. That is, it indicates that in natural things there is an inherent trend (or tendency) for substantial change as well as for change in quality, change in quantity and change of place. While keeping in mind that nothing moves or changes without the potential to change (except Aristotle's Prime Mover, which is pure act or actuality) this inherent trend or tendency to change (in all its forms) is a defining characteristic of natural things and so can be equated with *phūsis* itself.

Aristotle's *Physics*, particularly the second book, is concerned with what is moved, with extended matter, and particularly with 'what has the cause of its motion in itself'. In the second book of the *Physics* Aristotle argues that 'Nature' (*phūsis*) 'is the principle and cause of motion and rest to those things, and only those things, in which she inheres primarily, as distinct from accidentally'.[90] Here there is a strong parallel between Spinoza and Aristotle because Spinoza is concerned with what is moved and what has extension and because 'Nature' for each of them is not only a force acting from outside a thing, but also something intrinsic to the thing itself. In each there is a sense that this 'force' of 'Nature' acts within all things. Spinoza uses the term 'force' (*vis*) in different ways. For example at *Ethics* IIIP54D he says that 'the mind's striving, or power, is its very essence' and elsewhere uses the expression 'natural love' (*natuurlyke Liefde*) instead of *conatus*[91]. Spinoza also used 'force' as a synonym of

89 Zeller, Outlines of the History of Greek Philosophy, p. 178.

90 Physics B, 192b20 (Loeb edition), transl. Wicksteed and Cornford.

91 Wolfson, Philosophy of Spinoza, Vol. 2., p. 197.

'motion' in, for example, Proposition 22 of his *Descartes' 'Principles of Philosophy'* 'If body A is equal to body B, and A is moving twice as fast as B, the force, or motion, in A will be twice that of B.' In the demonstration of this he says '...by force in moving bodies, we understand a quantity of motion...But in bodies at rest we understand by 'force of resisting' a quantity of rest'. Here 'force' in the sense of (potential) motion has strong similarities to Aristotle's *dūnamis*. Further similarities are explored below, but it is worth noting in passing that the expression 'natural love' may be traced to Augustine's *amor*, the universal principle of self-preservation and to Thomas' *amor naturalis*, the natural tendency of bodies to move upward or downward. A sense of 'biological love' with a debt to these notions will be seen later to be important for contemporary environmental philosophy.

Two Levels of Conatus

For Spinoza, all particular entities have a *conatus* simply because they are extended substance. But is the *conatus* that, say, a rock has the same *conatus* (the same endeavour to persist in being), that a living organism has? The answer is yes, and no. There are two levels or categories of Spinoza's *conatus*—one category that all particular, individual things possess, and another category that all and only living particular things possess. Another way of saying this is that there is a psychic and a non-psychic dimension to Spinoza's *conatus* such that all things have 'non-psychic' *conatus* but only living things have a 'psychic' *conatus*.[92] This 'psychic' dimension should not, however, be interpreted in terms of modern psychology, but should be understood more in terms of the Greek concept of *psychē*, or animating essence.

To clarify further these categories of *conatus* it is worth going back to Gill's claim that Aristotle distinguished between an active and

92 Spinoza, Ethics IIIP28D.

a passive *dūnamis* . This distinction, between two sorts of immanent motive principle, may explain why Spinoza could recognize that living things had *conatus*, while also recognizing that something like a rock had *conatus*. In Aristotelian terms, the rock has a passive potential (*dūnamis*) to persist in its being, that is, it will persist as what it is unless it is acted upon by an external active *dūnamis* corresponding to its passive *dūnamis* . The rock has a potential to be acted upon. A living thing, on the other hand, has within it a passive and an active *dūnamis*. The passive *dūnamis* determines its response to forces applied to it. The internal *active dūnamis* that only living things have, allows a living thing to apply force to other things and so move them in some way, but, more importantly, this internal active *dūnamis* also allows a living thing to operate on itself to maintain itself, as itself. This internal active *dūnamis* is, in other words, an essential element of a living thing—an essential element of its nature because it is this that allows a living thing to maintain its identity through a process of change. From this it is clear that there is much in common between the concept of *conatus* and the immanent teleology of Aristotle that centres on his concept of *dūnamis*. Each attempts in its own way to refer to living individuals while also referring to the inanimate world; each attempts to explain what it is that allows a thing to persist as itself over time despite changes; and each refers to a force internal to the individual. This force, or dynamism, is not brought about by anything external to the individual—there is in each case an internal *telos* operating. The similarity between the 'psychic' sense of *conatus* that is inherent only in living things and Aristotle's *dūnamis* can be seen, for example, in the use of *phūsis* at *Metaphysics* Θ. Here, 'nature' (*phūsis*) is described as being in the same genus as *dūnamis* and as being an 'active principle of change...not in another thing but in the thing itself qua itself' (1049b9-10).[93] On this, Gill says that an 'entity's nature (*phūsis*) is an active *dūnamis* but... [it can also be that] ...an entity's nature is a source of change in the thing itself *qua*

93 Trans. Mary Louise Gill, Aristotle on Substance.

itself.'[94] That is, there is an active *dūnamis* bringing about change in an entity from within the entity. This active *dūnamis*, as the entity's *phūsis* or nature, would appear to be the same *phūsis* which, as discussed in the previous chapter, Mourelatos has characterized as a PHŪSIS or special sense of innate urge. The active *dūnamis* is an innate urge for change and the 'psychic' sense of Spinoza's *conatus* has a strong similarity to—if not identity with—Aristotle's active *dūnamis* or innate urge for change.

That Spinoza's *conatus* may be considered substantially the same as 'innate urge' in Aristotle's *phūsis* can further be seen by comparing Mourelatos' analysis of Aristotle's *dūnamis* and Spinoza's own account of *conatus*. Mourelatos argues that Aristotle meant that powers were located in things and that Aristotle had two levels of *dūnamis*: a general level where it is a 'principle (or source) of motion in something other than the thing moved or in the same thing *qua* other'; and more narrowly where powers are present in things not accidentally, but essentially. For Mourelatos this smaller group of powers in the narrower sense of *dūnamis* 'constitutes the essence of a thing, the sum of its essential powers [and] is its nature or *physis*'.[95] By comparison, Spinoza says that '...the power, or striving, by which it strives to persevere in its being is nothing but the given, or actual, essence of the thing itself'[96] and that 'the power [*potentia*] of each thing' is 'the striving [*conatus*] by which it does anything'.[97] For Spinoza the *conatus* of a thing is its essence and is the same as its *potentia*. *Potentia* is the common Latin translation of the Greek *dūnamis* so, taking into account that for Aristotle, in one sense at least, a thing's *dūnamis* is also its essence, there is a clear correlation between *dūnamis* and *conatus*.

94 Gill, Aristotle on Substance, p. 218.

95 Ibid., p. 99.

96 Ethics IIIP7 transl. Edwin Curley.

97 Spinoza, Ethics IIIP7D. The italicized word 'or' in the text is Curley's translation of the Latin *sive* or *seu* which he says generally indicates equivalence (Curley, Collected Works of Spinoza, p. xv).

That the concept of *conatus* in Spinoza, as the endeavour to persist in being, might ultimately owe its origins to Aristotle's *dūnamis* is also indicated by the definition of motion or change in Aristotle's *Physics* III.1, Wicksteed's translation of which reads:

> Reverting, therefore, to the universal distinction already established between 'being-at-the-goal' in actuality (*entelecheia*) and being in potentiality 'such-as-is-capable-of-attaining-the-goal' (*ē toū dunamei ontos entelecheia*), we can now define motion or change as the progress of the realizing of a potentiality, qua potentiality ... the actual growing or shrinking...of anything capable of expanding or contracting...the process of coming into existence or passing out of it...the actual moving of the physical body...[98]

Of particular note here is the distinction between being at the goal, which is a completed action, and the potential to get there, which is always an uncompleted process. The *dūnamis*, the potential, is never completed. The realization of *dūnamis* is *entelecheia*. Just as *dūnamis* is necessarily always incomplete, so too is *conatus* necessarily always incomplete, it is always the *striving* to persevere.

For Spinoza, just as for Aristotle, the active power of a living thing to maintain itself as itself is its essence or nature. But, as Mourelatos demonstrates, there is more than one sense of 'nature' employed by Aristotle and we should therefore also ask how Spinoza viewed 'nature'.

At least two senses of 'nature' can be distinguished in Spinoza — *Natura* (with an initial capital) which is the whole of nature (all existing things), and *natura*, which is essence. For Spinoza the essence (*natura*) of a thing is identical with the thing itself whereas *Natura* is further split into *Natura naturans* and *Natura naturata,* distinctions made in the *Short Treatise*. *Natura naturans* is 'a being that we conceive clearly and distinctly through itself [alone]...that is, God.'[99] Spinoza notes that the Thomists also used *Natura naturans* to refer to God, but that this was a being beyond all substance. *Natura*

98 Aristotle, Physics (201a 10-15).

99 Spinoza, Short Treatise, I(viii) in Curley ed. Spinoza Reader, p.57.

naturata is divided into a universal and a particular and 'requires some substances in order to be conceived properly'.[100] In the *Short Treatise* Spinoza says there are two universal *Natura naturata*: 'motion in matter and intellect in the thinking thing'.[101] This principle of motion within matter has parallels within Aristotle's *Physics* B.

Physics B and an Internal Principle of Movement

Aristotle says that the common feature that characterizes things constituted by nature is that they all 'have within themselves a principle of movement (or change) and rest'.[102] The phrase 'principle of movement (or change) and rest' translates *archen kinēseōs kai staseōs*. For Aristotle, the defining characteristic of things that come about through 'nature' (as opposed to the artifice that gives rise to, say, a bedstead)[103] is that they have some inherent principle for change or movement.[104] It is this inherent principle that has been the focus of this and the previous chapter and that will be used in the following chapters.

For Spinoza all natural things (i.e. in general terms, all extended things[105]) have an inherent principle of change or movement, a principle which operates specifically for the self-preservation of the thing with the principle. This principle, which Spinoza calls the *conatus* for self-preservation (striving to persist in being), inheres in

100 Curley, Spinoza Reader, p. 58.
101 Spinoza, Short Treatise I, ix.
102 Aristotle, Physics 192b11-14 (Loeb edition).
103 Ibid., 193a10-15.
104 Ibid., 192b14.
105 All natural things are extended things because 1) for Spinoza, there is only one substance and 2) this substance has infinite attributes, the only two of which we know being extension and thought.

all things and is the source of motion or change that all things have. All things have this principle even though it is manifested to varying degrees such that the principle generates movement or change (in all its senses) to a greater extent in an animal than a plant and in a plant more than a rock. Cicero and Diogenes Laertius limited *conatus* to animals, Augustine applied it to plants as well, but for Spinoza the striving to persist in being is in all things. So, for Spinoza, even a rock has the principle within it and is capable of change. Spinoza thought the main activities of bodies (as modes of the attribute of extension) are motion and rest, just as Aristotle did.

And just as Spinoza's immanent motive principle operates in all of extended nature, so too does Aristotle's motive principle operate in all of what is for him also extended nature (*phūsis*)—as can be seen in this passage from *Physics* B, 1.

> Some things exist, or come into existence, by nature; and some otherwise. Animals and their organs, plants, and the elementary substances—earth, fire, air, water—these and their likes we say exist by nature...and the common feature that characterizes them all seems to be that they have within themselves a principle of movement (or change) and rest (*archēn kinēseōs*)—in some cases local only, in others quantitative, as in growth and shrinkage, and in others again qualitative, in the way of modification...[106]

The phrase *archēn kinēseōs* is one that Heidegger, for example, emphasises in his treatment of the second book of Aristotle's *Physics.*[107] He says *phūsis* for Aristotle is *archē* [sic] *kinēseōs* ', the 'origin and ordering of change, such that each thing that changes has this ordering within itself.'[108] After outlining (in his own inimitable way) the four senses of movement, or change, that Aristotle employs, and exhorting his readers not to understand movement or change in a limited way (particularly not in terms of the commonly-understood

106 Physics 192b8-16.

107 Martin Heidegger, On the Being and Conception of PHYSIS in Aristotle's Physics B. 1, Trans. Thomas J. Sheehan, Man World 9 (1976): 219-276.

108 Ibid., p. 230.

movement implied in change of place), Heidegger argues that Aristotle wants to distinguish between the *archē* (origin) of natural things and the *archē* of manufactured things. Under this distinction the *archē* of natural things (e.g. growing things, living things but also non-living things such as rocks) is within themselves, whereas 'the *archē* of artefacts is *technē*)' or 'know-how'[109] which is applied externally. That is, the cause of artefacts is an external cause, not an internal cause. In the case of artefacts, of manufactured things, the originating principle that generates them is not in the artefact itself but in another, in the architect or designer. This reading of the *Physics* indicates that there is an originary principle of movement, of change, that is internal to all 'natural' things, and not just living natural things. 'Natural' here should be understood in both senses of the standard translation of *phūsis* just cited, i.e both with and without an initial capital.

Heidegger's translations of early Greek are, however, notoriously idiosyncratic and it would be prudent to seek confirmation elsewhere. Such confirmation is readily found in Collingwood, who looks to *Physics* Δ for clarification of Aristotle's use of *phūsis*.[110] There he finds seven possible meanings for the term which are one by one examined by Aristotle, then discarded, until the seventh meaning is reached (although it should be noted that the third meaning, the ordinary non-technical Greek usage as that which is the source of movement or change in natural objects, is not completely abandoned). The seventh meaning of *phūsis* offered by Aristotle is that it is 'the essence of things which have a source of movement in themselves'.[111] This Aristotle regards as the true and fundamental meaning of *phūsis*. Collingwood, like Heidegger, contrasts *phūsis* with *technē*, i.e. what things are when left to themselves as opposed to what human skill can make of them. From this contrast Collingwood determines that what Aristotle meant by the nature (*phūsis*) of things was the principle of

109 Ibid., p. 231.
110 R.G. Collingwood, The Idea of Nature, Oxford: Clarendon, 1945, pp. 80-85.
111 Ibid., p. 81.

growth, organization and movement that natural things have in their own right. When Collingwood's accounts of the third and seventh meanings offered by Aristotle are combined there is formed a definition of *phūsis* very close to that for which Heidegger argues.

These readings of the meaning of *phūsis* parallel Spinoza's view of *conatus* as covering all extended nature.[112] The *archēn kinēseōs* in Heidegger's interpretation (the principle of growth, organization and movement in Collingwood's treatment and the 'principle of change' or 'motive principle' in other translations) has much in common with Spinoza's notion of *conatus*. In Heidegger there can be found further confirmation that the concept of *conatus* is in essential respects implied in *archēn kinēseōs*. This occurs in his parenthetical comment that 'there is no reason to read *archē* in place of *hormē* in this text [i.e. *Physics* B,1] as Simplicius does, for *hormē*, 'impulse', elucidates the Being of *archē* quite well.'[113] As was noted earlier, the concept of *conatus* is derived, in part at least, from the Greek concept *hormē*.

But, to return to the relationship between *conatus* and *dūnamis*. In both Aristotle and Spinoza, Nature is that with an immanent motive principle: i.e. the motive principle needs to be immanent in some*thing*. Taking Blair's translation of *entelecheia* as that which has its end in itself,[114] 'nature' (*phūsis*) in Aristotle consists in those extended things that have their end in themselves. For Aristotle, this end-in-itself, this internal *telos* or inherent impetus toward some end is necessarily an inherent impetus for change existing in all (extended) nature. The end toward which the thing directs itself is, for Aristotle, the good of the thing itself—where that which is good for the thing is that which, by definition, acts to ensure the continuance or persistence of the thing. That which is good for a thing is that which allows it to persist in being, and so there is a clear relation between *conatus* and *entelecheia*. And, because there must be *dūnamis* (potential) for there to be *entelecheia* (actuality) or *energeia* (activity)—except in the case

112 Spinoza, Ethics IP34.
113 Heidegger, On the Being and Conception of... , p. 231.
114 Blair, Energeia and Entelecheia.

of the Prime Mover—there is a clear relation between *conatus* and *dūnamis*.

It is apparent that Spinoza drew significantly on Aristotelian concepts in the formation of his concept of *conatus* and that the single biggest influence was the concept of *dūnamis*, or potential. Spinoza's *conatus* inheres in all extended substance, just as Aristotle's *dūnamis* does. The difference in Spinoza's account is that there is, for him, only one substance, whereas, for Aristotle there are many.[115]

Spinoza's concept of *conatus* is, in all probability, derived from, and can certainly be expressed in terms of, Aristotle's *dūnamis*. The equivalence of *conatus* and *dūnamis*, coupled with fact that the concept of *conatus*, and cognate terms such as conation and conative force, has a significant role in environmental philosophy indicates that there is a strong case for at least an implicit Aristotelian basis for some key elements of environmental philosophy.

This idea is explored in the following chapters.

115 '... the answers to logical, metaphysical and ethical questions have their common origin in his view of individual substances acting out their identity in the world', Kosman, 'Aristotle's Definition of Motion,' p. 62

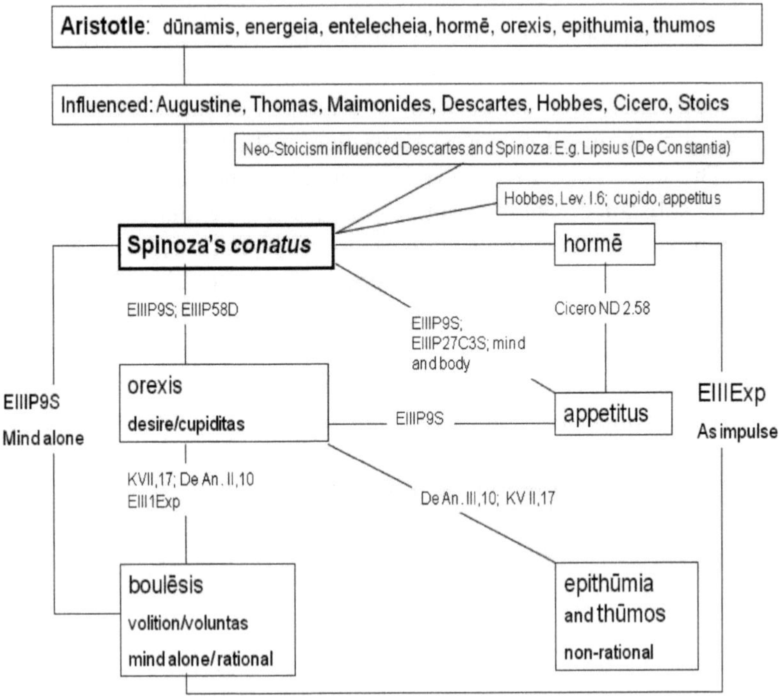

Figure 1: Influences on Spinoza's concept of *conatus*

PART II

MORAL CONSIDERABILITY

3. The Status of Organisms and Ecosystems

Introduction

From the discussion so far, it would appear that the quality in living things that sets them apart from other objects in the world can be described in terms of Aristotle's *dūnamis*, or in terms of at least one aspect of Spinoza's *conatus.* It is a quality that permits living things to determine their own goals—consciously or unconsciously—and defines them as autonomous entities.

There are several things this leads to, but before addressing these, it would be helpful to foreshadow an argument from the final chapters that living individuals are not mere existents but are bearers of a unique value that addresses an 'ought' to all moral agents. The unique value derives from each individual living thing having an internally-determined directiveness or purposiveness and so being an end-in-itself. As an end-in-itself, a living individual is also a good-in-itself because the individual in some sense values its own ends. This good-in-itself makes a claim to be protected and promoted and places any entity that recognizes the good under an obligation both to prevent it from being harmed and actively to support its flourishing. The prime example of this is a newborn baby. The baby is not a mere existent but an always already-valued existent that addresses an 'ought to care' to all moral agents. In the first instance the ought is addressed to the parents, but not only to them. The ought is addressed also to any being that recognizes the unique value, the immanent purposiveness, in the baby. To say it another way, the possession of the ability to determine (consciously or not) one's own goals (i.e. the possession of immanent purposiveness), coupled with the power to act on that determination makes all living things *prima facie* morally considerable and that simply being alive is a sufficient criterion of moral considerability.

This comes about in part because each living thing also has an *interest* in fulfilling its immanent *telos* and because it is within the power of a moral agent to affect that interest—whether by promoting it or hindering it.[1] It turns out that being alive is a sufficient criterion of moral considerability, but this is not to say that simply being alive is also a necessary criterion: rocks and mountains and wild places generally may have a claim to moral consideration on other grounds. It also needs to be made clear which things have the immanent *telos* that makes them *prima facie* morally considerable.

The phrase 'ability to determine one's own goals' implies that the entity able to do so is an individual, so the main question addressed in the first part of this chapter is whether *individual* organisms are the only entities possessing the immanent *telos* that would make them (*prima facie* at least) morally considerable. Having the property of being alive as the basis of moral considerability does not, however, necessarily mean that all organisms are equally morally considerable: a system for giving priority to one organism over another is a separate matter to the establishment of a basis for moral considerability—and one that is taken up in the final section of the book.

That individual organisms are morally considerable is an intuitively satisfying claim, but there are also significant arguments in favour of extending moral considerability beyond organisms, to ecosystems, for example, or even to inanimate natural features such as rocks and mountains. One argument for extending moral considerability to ecosystems is offered by Warwick Fox[2] under what he calls 'autopoietic ethics', in which all and only those things that maintain themselves through the process of autopoiesis are morally-considerable.

The term autopoiesis was created in 1972 by Humberto Maturana, then a research biologist at the University of Chile working on cognition as a biological phenomenon. He and his co-worker

1 The question of what constitutes an interest is important and will be addressed later.

2 Warwick Fox, Toward a Transpersonal Ecology, Boston: Shambhala, 1990.

Francisco Varela had been trying to identify and put a name to the autonomy which is a central feature of what they called 'the organization of the living,' that is, the central constitutive feature of life'. Maturana coined the word autopoiesis to describe the process of self-production by which an organism maintains itself as a unity even though all the component parts of the organism change over time. The early publications on autopoiesis argue for it to be applied primarily to single cells and only possibly to multi-cellular organisms. For Maturana and Varela the cell is the 'molecular embodiment of autopoiesis.'[3] Fleischaker[4] also takes this line, and her examples, like Maturana and Varela's, are all based on molecular biology. Fleischaker insists that autopoiesis can only be said to operate with certainty at the molecular level. This molecular level, the level of the individual cell, is what Kenny[5] calls first-order autopoiesis. If there is a second-order autopoiesis (which Fleischaker doubts) it is embodied in multi-celled organisms. Some suggest that there is also a third-order of autopoiesis in things such as ecosystems but the question of whether there is third order of autopoiesis is hotly contested. To be autopoietic something must be identifiable against a background and have identifiable boundaries which the unity itself determines using its own components. The components must come from internal transformation and have properties that are 'capable of satisfying certain relations that determine in the unity the interactions and transformations of these components.'[6]

3 Humberto Maturana and Francisco Varela, Autopoiesis and Cognition: The Realization of The Living, Boston/Dordrecht: Reidel, 1980, p. 90.

4 Gail Rainey Fleischaker, Autopoiesis: System Logic and Origins of Life, Diss., Boston University, 1988.

5 Vincent Kenny, 'On the subject of autopoiesis and its boundaries: does the subject matter?', International Journal of General Systems, Vol. 21, 1992, pp. 187-196.

6 Francisco Varela, Humberto Maturana and R. Uribe, Autopoiesis: The Organization of Living Systems, its Characterization and a Model, Biosystems 5, 1974, p. 192. Gail Rainey Fleischaker, Autopoiesis: System Logic and Origins of Life, Diss., Boston University, 1988.

The concept of autopoiesis is flawed, it offers no advantage over *dūnamis* or *conatus*[7] and it is highly problematic whether it can be applied at all beyond single-celled organisms, yet Fox's treatment still needs to be taken seriously. But, before examining Fox's position, questions concerning what it is to be an individual and what relations hold between individuals and the environments in which they persist need to be addressed. By using the concept of supervenience along with some of Andrew Brennan's[8] arguments, immanent teleology is shown to be a characteristic only of individual organisms and not of ecosystems. If immanent teleology is a characteristic only of individuals, then Fox can be shown to have made an invalid move in his treatment of autopoietic ethics when he assumes that ecosystems should be considered morally considerable on the basis that they are (third-order) autopoietic unities, but more on that later.

Parts and Wholes

Brennan argues that ecosystems or communities can not be considered as unitary objects (i.e. individuals) in the same way as biological organisms, but that, despite this, an organism's role within an ecosystem is an important part of the organism's being. He focuses on the idea that singular biological organisms have essential or 'basic' properties because they are members of a natural kind. In addition he maintains that the complete biological story of what a particular item is needs also to take into account the supervenient properties associated with the item's involvement in a community. Attributing properties to individuals on the basis that they are members of a

7 Stephan Millett, Autopoiesis and immanent teleology: toward an Aristotelian environmental ethic, Diss., Perth: Murdoch University, 1997.

8 Andrew Brennan, Thinking About Nature: An Investigation of Nature, Value and Ecology, London: Routledge, 1988.

natural kind is move often ascribed to Aristotle, although there is considerable criticism of such a move (see for example work by Balme, Hull or Sober). There is, however, no need to dwell on this point here because the main concern is with the concept of supervenience and how it may be employed to understand the (moral) relationship between organisms and the communities in which they live and on which they depend for survival. The concept of supervenience is important because it allows the role that a living thing plays in a community or ecosystem to be taken into account when making moral judgements. And it allows this role to be taken into account without imputing any ontological singularity to ecosystems or communities, as, for example, Mathews does with her notion of the ecological 'self'.[9] Supervenience and ontological singularity are difficult concepts on which there has been considerable debate and both stand in need of clarification.

For Brennan, one of the important problems for philosophy of ecology is to determine whether 'ecosystems are items that are more like trees, or items that are more like a sum of fairly unrelated parts'.[10] He suggests that at one extreme there is Tansley's view that ecosystems are 'the basic units of nature' within which everything else is nested and at the other extreme is the view that the concept of an ecosystem is a wilful imposition of 'pattern and structure where there is merely chance aggregation'.[11] To clarify the distinction between a singularity and an aggregate, Brennan gives the example of an oak tree in his garden and a compost heap standing next to it. The oak tree is, to him, a unified, integrated collection of parts whereas the oak-tree-plus-compost-heap is a non-unified part collection, it is a simple aggregate and not a unity. One key difference between the two is that only unified integrated part collections (such as the tree) are relevant to the functional status of components: 'the acorn is the fruit of the

9 Mathews, The Ecological Self.

10 Brennan , Thinking About Nature, p. 83.

11 Ibid., p. 48.

oak tree, not of the tree and the compost heap.'[12] Brennan recognizes the oak tree as a natural kind that has an 'essence' (or basic properties) whereas the oak-tree-plus-compost-heap is not a natural kind and does not have an essence. Both can be considered 'wholes' of a sort but only the oak tree can be considered a 'unified part collection'. The properties of non-unified part collections, and Brennan includes communities and ecosystems in this category, are supervenient on the basic properties of the component parts that make up the community or system. And if it is a community, as opposed to an ecosystem, that is under consideration the component parts will be singular, living organisms. But what does it mean to say that there is a relationship of supervenience between two things?

In a relationship of supervenience, properties of one kind, A, supervene upon those of another kind, B, when things are A in virtue of being B. Brennan says organisms have two sorts of properties: 'basic morphological or taxonomic properties' that we can use to classify them as biological kinds and 'functional properties' that may be seen in how the organism interacts within an ecological system. These functional properties are supervenient on their basic properties, which means that different organisms can fulfil the same role (have the same functional properties) in an ecosystem. However, 'a change in an organism's functional properties will mean a change either in its basic properties, or a change in its environing circumstances'.[13]

But there is no universal agreement on what is meant by supervenience and Brennan does not make entirely plain in which sense he is using it. Of supervenience, David Lewis says, 'The idea is simple and easy: we have supervenience when (and only when) there could be no difference of one sort without differences of another'.[14] Mental respects, for example, supervene on physical respects if and

12 Ibid., p.84.

13 Ibid., p. 112.

14 D.K. Lewis, On the Plurality of Worlds, Oxford: Basil Blackwell, 1986, p. 14. Cited by Brian P. McLaughlin, Varieties of Supervenience, Elias E. Savellos and Ümit D. Yalçin eds., Supervenience: New Essays, p. 17.

only if there can be no difference of a mental sort without some difference of a physical sort. Brian McLaughlin notes that supervenience has to do with sameness and difference in that 'A-respects supervene on B-respects if and only if exact similarity in B-respects excludes the possibility of difference in A-rcspccts'.[15] Following from this, the core idea of supervenience, McLaughlin says, comes down to the idea of *dependent-variation*. For Jaegwon Kim, on the other hand, there are three 'desiderata' of supervenience, *dependence, co-variance* and *non-reducibility*.[16] Kim sees co-variance as the crucial component which best helps to clarify what supervenience is. Under Kim's desiderata, supervenient properties *co-vary* with their subvenient or base properties; are *dependent* on or are determined by, their base properties; and are *not-reducible* to their base properties. A relation of supervenience in general is one of dependent, non-reductive co-variance.

In his brief history of the concept of supervenience Kim says that the present use of the term is continuous with Hare's use of it in *The Language of Morals* (1952) but notes that the concept of supervenience was also used early in this century by the 'emergentists' in the formulation of the concept of emergent evolution. The doctrine of emergence 'is the claim that when basic physicochemical processes achieve a certain level of complexity of an appropriate kind, genuinely novel characteristics, such as mentality, appear as 'emergent' qualities'[17]. For Beckermann, an emergent property 'is not only novel in the sense that it has never been

15 Brian P. McLaughlin, 'Varieties of Supervenience,' Savellos and Yalçin eds., Supervenience: New Essays, p. 17.

16 Jaegwon Kim, Supervenience and Mind, Cambridge: Cambridge University Press, 1993, p.140.

17 Ibid., p.134.

instantiated before, but also in the sense that it could never have been predicted in advance'.[18]

Kim asks whether there is an interpretation of covariance that is strong enough to sustain supervenience as a dependency relation but weak enough not to imply reducibility. He suggests that there are two types of covariance—weak and strong—and offers two definitions for each type. Kim's definitions run as follows. Let A and B be two sets of properties, where A is supervenient and B is subvenient:

> *Weak covariance I:* No possible world contains things, x and y, such that x and y are indiscernible in respect of properties in B ('B-indiscernible') and yet discernible in respect of properties in A ('A-indiscernible').
>
> *Weak covariance II*: Necessarily, if anything has property F in A, there exists a property G in B such that the thing has G, and everything that has G has F.
>
> *Strong covariance I*: For any objects x and y and any worlds w_i and w_j, if x in w_i is B-indiscernible from y in w_j (that is, x has in w_i precisely those B-properties that y has in w_j), then x in w_i is A-indiscernible from y in w_j.
>
> *Strong covariance II*: Necessarily, if anything has property F in A, there exists a property G in B such that thing has G, and *necessarily* everything with G has F.[19]

The addition of the second modal expression 'necessarily' in strong covariance II is what distinguishes it from weak covariance II and ensures that the G-F correlation holds across possible worlds. Kim notes that weak and strong covariance might readily be called weak and strong supervenience and that R.M. Hare and Donald Davidson have both come out in favour of *weak covariance II*—Hare in respect of moral properties supervening on natural properties and Davidson in

18 Ansgar Beckermann, 'Introduction', Ansgar Beckermann, Hans Flohr and Jaegwon Kim eds., Emergence or Reduction? Essays on the Prospects of Non-Reductive Physicalism, Berlin: Walter de Gruyter, 1992, p. 15.

19 Kim, Supervenience and Mind, p. 142.

terms of psychological properties supervening on physical properties.[20]

Although Brennan does not say so explicitly, it would appear that he has in mind a version of supervenience consistent with the definition of *weak covariance II*. For Brennan, ecological properties in this presently-existing world supervene on the natural properties of individual organisms. That is, ecological properties *depend* in some way on natural, physical properties—although it is not clear just exactly what that dependence relation is because, as Kim cautions, 'what must be added to covariation to yield dependence is an interesting, and metaphysically deep question'.[21] Although it is not clear what the dependence relation is between individual organisms and the ecosystems they inhabit, it is likely to be a variety of part-whole dependence (mereological supervenience), which Kim says may constitute a special and basic category of dependence.[22]

Before continuing with Brennan's position, it should be noted that supervenience theses generally can be grouped together as attempts to describe a non-reductive form of physicalism, something that autopoiesis can also be said to attempt. Autopoiesis is addressed here because the claims made of it are grand and too-ready acceptance of it raises the risk of effacing a rich understanding of the moral status of organisms. The theory of autopoiesis holds that the cognitive states and properties that define a system as living are 'realized' (made real) by the physical states of the systems of which they are properties, that autopoiesis describes the 'realization of the living'. Realization, according to Beckermann, is the 'complement of emergence' in which if a state or property of a system (S) is 'realized by a physical state of S, it is not emergent'.[23] The position represented in the concept of autopoiesis, in effect denies that cognitive states are emergent

20 Ibid., p. 142.

21 Ibid., p. 148.

22 Kim says mereological supervenience may constitute a special and basic category of dependence. Ibid., p. 148.

23 Beckermann, 'Introduction', p. 18.

properties. Because the cognitive states and properties of an autopoietic system are not reducible to the physical states of the system, the theory of autopoiesis should be considered a variety of non-reductive physicalism and if it is seen in this way then the relationship between the whole living thing and its component parts may be explained in terms of mereological supervenience. And if autopoiesis were to be extended to a third level, to the level of groups of organisms or systems of living individuals, then this level of autopoiesis also could be explained in terms of supervenience. But there are some significant problems with the third level of autopoiesis in general and especially assumptions that such a level of autopoiesis applies to ecosystems and justifies their moral considerability being addressed as if they were unities.

The dependence relation between organism (as part) and ecosystem (as whole) is at the heart of Brennan's discussion—a discussion that relies in part on arguing against the notion of 'ontological parasitism'. To make this argument Brennan looks to the desk on which he is writing and says that if what is supporting his keyboard is a collection of desk parts then the concept of the desk itself is 'an ontological parasite on its parts'. But, he says, it would be wrong to think of the desk as *merely* a collection of parts because, although it is a collection of parts, the parts are organized and structured into a particular shape: the parts are causally-integrated. The desk is a unitary object such that if one of its parts is interfered with in certain ways, the other parts are also affected. It is absurd, says Brennan,

> to maintain that what is really there is simply that material upon which we work to produce artefacts and that artefacts are not themselves anything but parasites on this material. For the claim that artefacts like my desk are *real* is not the claim that in manufacturing such things we add to the basic stock of *materials* in the world.[24]

24 Brennan, Thinking About Nature, p. 81.

He suggests that the difference between unified wholes and mere sums of parts is not directly linked to the question of reality and that wholes are just as real as their parts, although they are different.

There are implicit in Brennan's argument a number of ideas that could be described as Aristotelian—his view of essence, for example. His argument becomes explicitly Aristotelian when he tries to find a workable answer to the difficult question: what are the features of a unified thing of a (natural) kind? In trying to answer this question he merges ideas from Aristotle and Locke and adds to them Shoemaker's idea that there are two levels of causal powers in objects.[25] From Aristotle he takes the idea that the properties of things are constitutive of their nature and in some strong sense responsible for the thing being the kind of thing it is. From Locke he takes the suggestion that properties—what Locke called 'qualities'—are powers that produce ideas in us. The combination of these leads him to suggest that genuine properties can be related to causal powers possessed by objects. Under Shoemaker's distinction between higher and lower powers, genuine properties of objects can be considered as higher-level powers that are definitive of the thing itself. Higher-level powers have the capacity to produce lower-level powers that are not constitutive of the thing (a car, for example, has the power to crush small mammal, but this power is not constitutive of the car). This distinction between higher and lower powers is reminiscent of the distinction Mourelatos draws between different levels of *phūsis* and of *dūnamis* in Aristotle—although Mourelatos' distinction is not obviously the same as Shoemaker's. In Mourelatos' terms a living thing has the power to maintain itself as itself as well as the power to affect others. A living thing's powers of self-maintenance are essential to its being whereas its power to affect another depends also on that other having a passive 'power' (*dūnamis*) to be affected. To re-cast Brennan's example in Aristotelian terms: an elephant has the power to

25 Sydney Shoemaker, Identity, Cause and Mind, Cambridge: Cambridge University Press, 1984. Cited by Brennan.

maintain itself as itself and incidentally has the power to crush a small mammal. The elephant's ability to crush the mammal, however, requires also that it be in the mammal's essential nature to be crushable.

In a move similar to that of Mourelatos' analysis of Aristotle's powers, Brennan maintains that properties are *clusters of powers* and that this 'gives us more or less what we need in order to maintain the thesis that unified, causally integrated whole objects are precisely those that have *genuine* properties *in their own right...* this is to say they have powers and natures, and belong to sorts or kinds. The sum of the tree and the compost heap [in Brennan's garden] has no powers in its own right'[26] whereas the tree alone does. There are strong similarities between this and the position ascribed to Aristotle in which only individual living things, which are members of a natural kind, have certain properties in their own right. Things that are merely a sum of parts have no powers in their own right, they have only those powers that they inherit from their parts.

Brennan does not deny that certain wholes are more than merely the sum of their parts, but says that ecosystems and communities are not among those wholes that can be categorized in this way. If ecosystems and communities cannot be categorized as wholes they cannot be considered to be third-order autopoietic unities and Fox's inclusion of Gaian ethics under a more general class which he calls autopoietic ethics fails.[27] 'If there are such things as ecosystems and biological communities,' Brennan says, 'then their properties will be explicable by reference to the properties and interactions of their parts'.[28] His argument again becomes explicitly Aristotelian when he denies that ecosystems or the climax communities at the end of an ecological succession are an end or final cause in Aristotle's sense, whereas it is obvious to 'those of an Aristotelian frame of mind...that each living thing has certain aims, goals or ends towards which it

26 Brennan, Thinking About Nature, p. 86.
27 Fox, Toward a Transpersonal Ecology.
28 Brennan, Thinking About Nature, p. 88.

strives'.[29] These aims, goals or ends are determined from within the individual.

From the Aristotelian perspective argued for in the preceding chapters, there is interaction and a relationship of reliance between the elements that make up an ecosystem, but the ecosystem or community itself has no *internal* principle of unity. There is no active principle of internal organization holding the ecosystem together, it is only an aggregate of parts and should not be considered a unity. The ecosystem does not have a unique *dūnamis*. Nor does it have its own *conatus*: though each of its component parts has. And, in the same way, an ecosystem should not be considered an autopoietic unity.

Brennan recognizes that his approach to essence and supervenience is in opposition to, for instance, Quine, who wants to avoid the 'metaphysical jungle of Aristotelian essentialism',[30] but he says it does not matter whether the facts of an organism's ecological role are considered essential or merely contingent. The ecological role of the organism supervenes on characteristics that are internal to the organism. Even if it is necessary for the purposes of scientific study to separate ecosystems or communities from others, it is still the organism that is the *basic* unit of considerability because it is the organism that has an immanent *telos* that it has a vital interest in pursuing. Neither ecosystems nor communities have this same immanent *telos*: they may be worth protecting, but they are worth protecting primarily because doing so protects the interests of the individuals which comprise them. In protecting individuals and in ensuring there is a variety of species the sum of value in the world is protected.

The claim that it is the organism which is the basic unit of considerability because it has immanent *telos*—is, on the face of it, not without problems. For instance, it is at odds with what is often characterised as a Kantian position—which holds that only rational

29 Ibid., p. 109.
30 Ibid., pp. 116-117.

beings are morally considerable. More, then, needs to be said in its support.

Ends-in-themselves

In Kantian ethics ends-in-themselves are the foundation of morality—there is nothing more fundamental from which morality can be derived. But in choosing to make the possession of rationality a defining characteristic of ends-in-themselves,[31] Kant adopts a too-narrow view of what an end-in-itself is. This is not to take issue with Kant's more general position that those beings which are ends in themselves are *prima facie* morally considerable. However, it is to argue that all living individuals, and not just those that possess rationality, are ends-in-themselves and so are morally considerable. In short, there is a need to question what it means for something to be an end in itself and, because of the thoroughgoing influence Kant's work has had on this question, Kant is a suitable place to begin.

Suppose, says Kant,

> there were something *whose existence* has *in itself* an absolute value, something which as *an end in itself* could be a ground of determinate laws; then in it, and in it alone, would there be the ground of a possible categorical imperative—that is, of a practical law.[32]

Now, this in itself does not limit the class of morally considerable entities to those that possess something called 'rationality'. Kant, however, adds that

31 See Immanuel Kant, Groundwork of the Metaphysic of Morals, 3rd ed., transl. H.J. Paton, New York: Harper, 1964, p. 95.

32 Ibid.

> man, and in general every rational being, *exists* as an end in himself, *not merely as a means* for use by this or that will.[33]

He adds further that non-rational beings (what he calls 'things') have only a relative value and not the objective value that rational beings (what he calls 'persons') have.[34] But why? Kant's argument is that, if there is to be a supreme principle on which is based a categorical imperative

> it must be such that from the idea of something which is necessarily an end for every one because it is an *end in itself* it forms an *objective* principle of the will and consequently can serve as a practical law. The ground of this principle is: *Rational nature exists as an end in itself.* This is the way in which a man necessarily conceives his own existence: it is therefore so far a *subjective* principle of human actions.[35]

But neither rational nature being an end in itself nor man conceiving his existence in a certain way precludes other existing things from being ends in themselves.

Kant held that there are two kinds of ends, *producible ends* and *self-subsistent ends*. Producible ends are ends that agents try to bring about, whereas self-subsistent ends are existing beings for whose sake producible ends are produced. According to Donagan, 'producible ends are necessarily conditional upon the existence of self-subsistent ends, and on what the needs of these ends may be. Self-subsistent ends, by contrast, are ends-in-themselves.'[36] Donagan agrees with Kant that these ends-in-themselves are the foundation of morality, there being nothing more fundamental from which morality can be

33 Ibid.

34 Ibid., p. 96.

35 Ibid.

36 Alan Donagan, Morality, Property and Slavery, J.E. Malpas ed., The Philosophical Papers of Alan Donagan, Vol. 2, Chicago: University of Chicago Press, 1994, p. 111.

derived.[37] But there does not appear in this to be any requirement for self-subsisting ends to be rational beings.

The distinction between producible and self-subsistent ends is one drawn by Thomas Aquinas and by John Duns Scotus. As Donagan relates,

> Scotus pointed out that the produced end of an action—he called it the action's *finitum*, what is finished or completed by it—is always something done out of love for something already existing, and he maintained that only the loved thing already existing can strictly be called the action's *finis*—its true end (or *telos* or Zweck). Kant calls the already existing things out of love for which it is rational to act 'ends in themselves', presumably to bring out that they are ends, not by virtue of being loved, but by virtue of being such that it would be irrational not to take each of them into equal account in doing anything for anybody's sake.[38]

These ends in themselves are for Donagan, as they are for Kant, 'every rational being'.[39] But why should rationality be a necessary property of something that is an end in itself? *All living things* have an internal *telos* toward which they will normally proceed and, as will be explained shortly, they have vital or welfare interests as a consequence. That is, they have immanently-determined ends and are thereby ends-in-themselves. What happens to each individual living thing *matters* to that thing—and a living thing does not have to be rational for the achievement of immanent ends to matter to it. From arguments centring on Aristotelian immanent *telos* raised earlier, the immanent end of all living things is to *be* an embodiment of a mature example of their kind, whatever that kind might be.

37 Ibid., p. 112.

38 Alan Donagan, The Moral Theory Almost Nobody Knows: Kant's, J.E. Malpas ed. The Philosophical Papers of Alan Donagan, Vol. 2, p. 149. In this paper Donagan says the distinction was drawn by Scotus. In 'Philosophical Ethics: What It Is and What It Should Be,' later in the same collection, Donagan notes that Aquinas and Scotus both drew the distinction. He does not say whether Kant was aware of Scotus' distinction.

39 Kant, Groundwork to the Metaphysic of Morals, p. 64. Cited by Donagan, 'Morality, Property and Slavery,' p.111.

Donagan attempts to explain why rationality is necessary for a being to be an end in itself. He says

> the heart of Kant's moral theory is his brief statement of why the being of each rational being is an end in itself. It is because all rational beings necessarily so regard themselves. By what necessity do they [so regard themselves]? Kant's answer is implicit. He invites his readers to reflect on these lines. Must not all beings who can act or not act according as they think best, distinguish themselves from beings that cannot as requiring different treatment.[40]

Yes, rational beings are distinct from non-rational beings. Yes, rational beings are ends in themselves. Yes, rational beings are autonomous, that is, 'capable of asking themselves what existing beings it is rational to act for the sake of...'[41] But there is nothing obvious in this which shows that *only* rational beings are ends in themselves—at least ends in themselves as outlined earlier. Both Kant and Donagan hold that only creatures that have a capacity for self-reflection can be ends in themselves. But, what is so important about this capacity? Why is this capacity *necessary* when there is immanent purposiveness in all living individuals—an immanent purposiveness that the individual need not be aware of but which, at a very basic level, makes the individual self-valuing.

Kant contends that there can be no requirement to promote a thing's good unless that thing is an end in itself—'an existing being whose 'being there' (*Dasein*) is not merely something to be used for the sake of other things'[42] As far as this goes, there is no problem. The problem comes with the argument that only those beings that are rational are ends in themselves. What is the critical importance of rationality? It is almost certainly a defining characteristic of a moral *agent*, but equally certainly is not a necessary or defining characteristic of a moral *subject* or patient—where a moral patient is

40 Alan Donagan, Philosophical Ethics: What It Is and What It Should Be, J.E. Malpas ed., The Philosophical Papers of Alan Donagan, Vol. 2, p. 277.

41 Donagan, The Moral Theory Almost Nobody Knows, p. 149.

42 Donagan, Philosophical Ethics, p. 277.

an entity with vital or welfare interests that can be harmed or benefited. All moral agents may themselves be morally considerable on the grounds of their rationality, but although moral agency may be a sufficient condition for moral considerability, it is not a necessary condition.

All living things are ends in themselves and are *prima facie* moral patients because they have interests that can be harmed or benefited. All rational beings are ends in themselves, but they are also *prima facie* moral agents. Moral agents may well have a claim to greater consideration because they have within themselves the power consciously to affect the welfare of other beings, but it cannot with justification be claimed that moral agents are themselves the only objects of moral consideration.

Ends in themselves are said to have intrinsic value as well as interests that can be harmed or benefited. These concepts of intrinsic value and interests are, however, frequently ill-defined and some clarification is needed.

Intrinsic Value

Generally, environmental axiologies (theories of value) take either an anthropocentric and instrumental view of nature—in which the natural world has value to the extent that it has value for humans—or else they assert that there is some kind of intrinsic value for the natural world and/or the things that dwell in it. Fox[43] gives useful summaries of both instrumental and intrinsic value approaches to environmental ethics

Within these two broad divisions there are many different species of environmental axiology. It is, however, the idea of intrinsic value

43 Fox , Toward a Transpersonal Ecology, pp. 150-161

and its necessity or otherwise for an ecocentric worldview (one in which humanity and 'nature' are, phenomenologically speaking, one) and ethics that has been the focus of much recent environmental philosophy and that is of greatest interest for the discussion here.

The term 'intrinsic value' has a rich history in recent environmental literature; including work by Arne Naess, Richard Sylvan, Val Plumwood, Freya Mathews, Bill Devall and George Sessions, Paul Taylor, Kenneth Goodpaster, J. Baird Callicott, Tom Regan and Holmes Rolston III. Of these, Callicott and Regan, for example, both contend that a proper environmental ethics requires that inherent, or intrinsic, value in nature be postulated.[44] Brennan, however, disagrees with the view that a notion of intrinsic value is necessary[45] and is, in Rolston's words, 'agnostic' on the existence of intrinsic value.[46] Mathews' summary of intrinsic value[47] indicates clearly that there is no uniform approach to defining intrinsic value and so it is necessary to clarify what is meant by the term.

One of the most enduring attempts to clarify the concept of intrinsic value was by George Moore early in the 20th Century. In his treatment, Moore distinguished between the intrinsic *value* and the intrinsic *nature* of a thing so that,

> to say that a kind of value is 'intrinsic' means merely that the question whether a thing possesses it, and to what degree it possesses it, depends solely on the intrinsic nature of the thing in question.[48]

44 See, for example, J. Baird Callicott, Intrinsic Value, Quantum Theory, and Environmental Ethics, Environmental Ethics 7, 1985; Tom Regan, The Nature and Possibility of an Environmental Ethic, Environmental Ethics 3,1981.

45 Brennan, Thinking About Nature.

46 Holmes Rolston III, Review of Andrew Brennan's Thinking About Nature, Environmental Ethics, Vol. 11, Fall 1989, pp. 259-267.

47 Freya Mathews, The Ecological Self, pp. 174-180.

48 George E. Moore, The Conception of Intrinsic Value, Principia Ethica rev. ed., Edited and introduction by Thomas Baldwin, Cambridge: Cambridge University Press, 1993, p. 286.

When Moore says that the possession of intrinsic value 'depends solely on the intrinsic nature of the thing in question' he means to say two things, that:

> ...it is *impossible* for what is strictly *one and the same thing* to possess that kind of value at one time...and *not* to possess it at another; and equally impossible for it to possess it in one degree at one time, or in one set of circumstances, and to possess it in a different degree at another, or in a different set; and ... it is *impossible* that of two exactly similar things one should possess it and the other not, or that one should possess it in one degree and the other in a different one.[49]

For Moore, intrinsic value is an objective value—a value predicated solely on intrinsic properties. Moore's definition, however, is broader than that usually employed in environmental philosophy. Mathews, for example, defines intrinsic value as 'the value that beings have in themselves, regardlesss of whether or not they are valued by external valuers'.[50] She notes that what she calls 'intrinsic value' is in many respects the same as what Paul Taylor calls 'inherent worth'. She restricts the possession of intrinsic value to what she calls 'beings', and, secondly, and holds that beings value themselves (even if such valuation is not a conscious act)—that is, beings have intrinsic value at least because they are capable, in some sense, of valuing themselves. But what might a non-conscious valuing consist in? In answering this there is some assistance available from Aristotle's immanent teleology. From this perspective intrinsic value is a value possessed by an entity that has a 'good' of its own and which this entity possesses simply in virtue of it existing. In this definition an entity with a good of its own has an end toward which it can direct its endeavours (whether consciously or not). This would mean, for example, that any entity that has an interest in its own flourishing has intrinsic value. The 'good of its own' is something that a being pursues simply through the process of being or existing—the pursuit

49 Ibid., p. 286.
50 Mathews, Ecological Self, p. 178.

of this 'good' is what constitutes an entity as a (living) being and differentiates it from other objects in the world. Living things, and only living things, have an interest in pursuing their unique good.

Interest

The question of what an 'interest' is arises because of the need to determine criteria for moral considerability—that is, through the need to determine what are those characteristics or properties that a thing has by virtue of which it stakes a claim to be taken into account when moral judgments are made.

Questions of interests are often connected to the concept of 'harm'. Joel Feinberg, for example, says it is common, 'to take the object of harm always to be an *interest,*'[51] where the object of interest 'is what is truly good for a person' and where the inflicting of harm is conceived in terms of the violation of a person's interests, that is, in terms of an injury to something in which the person has a genuine stake. His discussion of harm centres on a 'harm principle' under which harm may only be justified if it is necessary to prevent even greater evils. For Feinberg harm is damage to one's *vital* interests, this damage occurring through a lack of something, or the presence of something, as well as through something being done or not being done. Harm, or damage to vital interests, may also be done to animals but Feinberg claims that humans may be justified in controlling animal movements, exploiting animal labour and even deliberately slaughtering animals. All these actions would be considered harms if inflicted on humans but Feinberg limits the harm which animals can suffer to invasion *only* of the interest that animals have in being free from pain that is inflicted cruelly or wantonly. There is little in favour

51 Joel Feinberg, Social Philosophy, Englewood Cliffs, NJ: Prentice Hall, 1973, p. 26.

of limiting the concept of harm in this way as each and every living thing has a vital interest that can be benefited or harmed—whether or not they are aware of the benefit or harm. Limiting harm thus, would seem to be merely a justification for some existing practices and would find serious opposition in, for example, Peter Singer's work on animal liberation or in Stephen R.L. Clark's work. Clark notes, for example, that 'animals in varying degrees feel pain and fear, and that we have no general reason to think them less subject to these ills than we' and that 'one should not cause unnecessary suffering to animals'[52] Feinberg usefully distinguishes between 'harm' and 'hurt', where hurt is a harm that the harmed party notices or is aware of. Something (or, more usually for Feinberg, someone) may be harmed without knowing it, but one can only be hurt if one knows about the harm done to oneself. Not all harms *hurt* because not all harms come to be noticed. All living things can be harmed, but not all may be hurt.

The concept of harm is taken up by Kleinig,[53] who also talks of harm in terms of violation of or intrusion on interests. Kleinig notes Feinberg's distinction between harm and hurt and acknowledges that it may be possible for even inanimate objects to be harmed, but he sees no advantage in explaining harm in terms of loss, damage, or injury, preferring to characterize harm as an interference with or the invasion of a person's interests. He recognizes that the concept of 'interests' is ambiguous and suggests three uses, not all of which would normally be related to the concept of harm: i) a psychological

52 Stephen R.L. Clark, The Moral Status of Animals, Oxford: Clarendon Press, 1977, pp. 42-3. See also, Stephen R.L. Clark, Animals and Their Moral Standing, London: Routledge, 1997. In 'The rights of wild things' in this volume he says there that even if humans have certain rights 'it seems difficult to see any plausible reason why non-human animals might not also have them.' (p.20) In 'Ethical problems in animal welfare' in the same volume, he describes animal welfare as 'that state or condition or habitual activity in which animals are well off', where being 'well off' is not merely a situation described by the absence of pain (p.114).

53 John Kleinig, Crime and the Concept of Harm, American Philosophical Quarterly, Vol. 15, No. 1, January 1978.

sense—unlikely to be related to this sense of interest—captured by the expression 'X is interested in Y'; ii) a sense in which my interests are restricted to projects in which I have a stake—in which I stand to gain or lose something—a sense generally expressed in statements of the form 'X has an interest in Y'; and iii) a sense of interest referred to by Kleinig as given under the form 'Y is in X's interests'. The interests at issue in this third sense are *welfare* interests where a being's welfare consists, most generally, in the absence of defects and irregularities with respect to some conception of its normal functioning. The *welfare* sense of interests is the most important in relation to the notion of harm at issue here.

In the case of animals (and perhaps plants), Kleinig suggests, welfare is largely a matter of being free of disease and deformity. He goes on to say:

> welfare interests, in their foundational aspect, are those interests which are indispensable to the pursuit and fulfilment of characteristically human interests...Being foundational, their satisfaction is not to be identified with a person's happiness or well-being so much as the conditions which make happiness or well-being possible.[54]

Although this refers only to humans, Kleinig notes later that harm may be done to animals as well as people and that any being with welfare interests can have harm done to it. Harm, he concludes, should be understood as the impairment of a being's welfare interests.

But is the impairment of a welfare interest morally significant? Consider the following example from Andrew Brennan: A crab is prodding at a mine buried in the sand. It is doubtful that the crab is aware of itself, even though it is aware enough not to try to eat its own claws, and aware enough to be prodding at the mine in a directed way and not just randomly stirring up the sand. In such a case the crab might be said to have a preference interest in exploring or moving the mine. But, if it triggers the mine to explode this action will be contrary

54 Ibid., p. 31.

to its welfare interests. One might conclude from this and similar examples that it is also preference interests, and not merely welfare interests, that are morally relevant.

Preference interests are morally relevant—but not at the expense of welfare interests. Welfare interests are morally relevant because they are basic interests that, if they are not satisfied, deny the possibility of any other interests being satisfied. In the above example the crab has a preference interest in not being prevented from stirring up the sand beside the mine. However, if the (basic) welfare interests of the crab in the above example were not satisfied it would die and so would be permanently prevented from exercising any preference. The key point here is that in satisfying welfare interests, one is preserving the *potential* for a being to satisfy any other interests it may have. Preservation of such potential is something that moral agents must take into account when making decisions that will affect moral subjects. However, whether the issue is preference interests or welfare interests, any extension of interests to non-humans is not uncontroversial. Many of the arguments for denying that non-humans have interests rely on the distinction that Williams[55] makes between 'having an interest' and 'taking an interest'. In Williams' sense, to *have* an interest in something it is sufficient that the 'something' may 'impinge on one's well-being,' so Williams' sense of interest is, at least in part, a welfare interest. To *take* an interest, however, requires an epistemic relationship to one's world because 'it requires an understanding of what one's interest is and what stands in one's interest'.[56] To *take* an interest would seem, then, to be much the same as what Brennan calls a preference interest—although it is not clear whether taking an interest, in Williams' sense, requires self-

55 Meredith Williams, Rights, Interests and Moral Equality, Environmental Ethics, Vol. 2, 1980, pp. 149-161.

56 Steve F. Sapontzis, 'The Moral Significance of Interests,' Environmental Ethics, Vol. 4, Winter 1982, pp. 345-358, p. 347, citing Williams, 'Rights...', pp. 152-156.

consciousness. Some, such as Frey,[57] argue that only those entities that can *take* an interest should be morally considerable. However, it is false to suggest that only humans stand in an epistemic relation to their world and take an interest in what is in their interest. This is a position put by Clark[58] and with which Johnson, for example, agrees. Some support for non-human epistemic relations with the world can be found in work with the higher primates but this has been strongly contested.[59] The question, though, is not whether standing in a *self-conscious* epistemic relation to the world, as humans do and as some, at least, of the higher mammals appear to do, is pertinent. The question is whether standing in an epistemic relation to the world in general is pertinent. That is, the key would seem to be whether an entity is in a learning relation with the world. Living things respond to the world around them in ways categorically distinct from any changes that may occur to inanimate entities. Living things have a capacity to respond to their world and also have a capacity for changing themselves. In changing themselves they are, in effect defining themselves—something that non-living entities cannot do. A living thing is a being with self-identity and the capacity to be harmed or benefited that is independent of how humans might regard that being.[60]

On the question of the basis of moral considerability, Kenneth Goodpaster has argued that because trees and other plants have an

57 R.G. Frey, Interests and Rights: the Case Against Animals, Oxford: Clarendon, 1980 cited in Lawrence E. Johnson, A Morally Deep World: an essay on moral significance and environmental ethics, Cambridge: Cambridge University Press, 1991.

58 Clark, Stephen R.L, The Moral Status of Animals.

59 See e.g. R. Allen Gardiner and Beatrice Gardiner, Teaching Sign-Language to a Chimpanzee, Science 165, 1969, pp. 664-672. It should be noted, though, that the Gardiners' conclusions from their work with chimpanzees such as Washoe have been strongly challenged on the basis that it demonstrates learned behaviour rather than language acquisition (and hence does not indicate an epistemic relationship to the world). See also the Great Ape Project.

60 Johnson, A Morally Deep World, p. 79.

obvious capacity to maintain themselves and to heal themselves if injured it is difficult not to accept that they have interests.[61] For Goodpaster, all living things *have* interests and the sufficient condition for moral considerability is the condition of *being alive*. It should be noted, however, that what constitutes 'being alive' is not always clear. What, for example, would be the moral status of a dormant seed? And would its moral status be different if it were the last seed of a now-extinct species—such as some of those kept in storage at Kew Gardens? Is such a seed morally considerable because it is actually alive, potentially alive or is it its status as a conservator of a species which is most morally relevant? In such a case is there a single entity in which value inheres and that can be said to have intrinsic value? These questions can be addressed by recasting interests in Aristotelian terms. In such terms the seed is *potentially* the mature form of its species—is potentially a specimen of the tree which originally set the seed. In effect the seed *is* the tree, potentially. The seed has within it an active *dūnamis* to act upon itself so that in the presence of certain background conditions it will become what potentially it is. In this way it can be seen that the seed itself has intrinsic value because of its *potential* to be a mature form of its species. The seed also has a value determined by its potential to affect other specimens of other species in a given environment. If the seed were to grow to maturity in a certain biotic community it would be an agent of change in that community (as all components of a community to a greater or lesser extent are): it would have a value to the community that supervened on any intrinsic value or value it had to itself. The supervenient value would not necessarily be wholly positive or negative: a tree might be a home to birds and other animals, but also shade out understory plants.

If we take the position that ecosystemic or community value supervenes on individual intrinsic value then ecosystems are not in themselves morally considerable because they are not autonomous

61 Kenneth Goodpaster, On Being Morally Considerable, Michael Zimmerman et al. eds., Environmental Philosophy, Englewood Cliffs, NJ: Prentice Hall, 1993.

entities with either a good of their own or interests (welfare or preference).

Intrinsic Value Theories in Environmental Philosophy

Fox discusses the main approaches to intrinsic value theory in environmental ethics under four headings, which is a convenient categorisation to start with. His four headings are: 1) ethical sentientism; 2) biological ethics and autopoietic ethics (which he says is often referred to misleadingly as conative ethics); 3) ecosystem ethics and ecosphere (or Gaian ethics); and 4) cosmic purpose ethics. He maintains that although they are often treated separately, ecosystem ethics and ecosphere ethics can be subsumed under autopoietic ethics, so it will pay to give a brief summary of the first and the last categories before returning to look at the middle two together.

Ethical Sentientism

Advocates of ethical sentientism, the most prominent of whom is probably Peter Singer, propose sentience (defined, for the present purpose, as the capacity for feeling or suffering) as the criterion of moral considerability. Sentientism generally takes the view that all sentient entities have interests and that they have intrinsic moral value. However, sentience in environmental philosophy has taken on a narrower definition than that originally applying to the term. Sentience is properly consciousness of something or the other, and not specifically pleasure or pain, but the idea that sentient beings seek pleasurable states and avoid unpleasurable states, and that they can suffer, has become widely accepted within environmental theories. One difficulty in this view is the circularity in the concept of seeking a pleasurable state of being because of the assumption that, in the case of rational sentient beings, any state of being that an entity seeks is a pleasurable state. If the state is not pleasurable then it is not rational to seek it. Another of the problems is that of where to draw the line between what is considered sentient and what is not, or, put another way, to decide what can suffer and what cannot. In the first edition of *Animal Liberation*, Singer says the line might be drawn for practical purposes between arthropods (lobsters, shrimps) and molluscs (oysters, mussels), it therefore being justifiable to eat molluscs but not arthropods. In the second edition of *Animal Liberation*, however, Singer says that while no-one can say with confidence that molluscs do feel pain, neither can anyone say with confidence that they do not. If it turns out that molluscs can in fact feel pain then eating a meal of oysters or mussels would inflict pain on 'a considerable number of creatures'. On this basis Singer concludes that it is better in moral terms to avoid eating molluscs.[62]

Within ethical sentientism, being sentient is deemed to be the sole criterion for the possession of interests. That is, only sentient entities can be benefited or harmed. However, being sentient may be sufficient for an entity to have interests, but is it necessary and is being sentient sufficient for an entity to be morally considerable? Being rational and being able to feel pain or pleasure might be sufficient conditions for moral considerability, but they are not necessary conditions. Goodpaster, for example, argues for an expansion of the criteria of moral considerability so that simply being alive is sufficient[63]. But this assumes that the condition of 'being alive' can be determined unproblematically—an assumption that proponents of the idea of autopoiesis, at least, would question, especially in so far as they have attempted to expand the applicability of autopoiesis beyond its origins in molecular biology.

62 Peter Singer, Animal Liberation 2nd ed, New York: Harper Collins, 1990, p. 174.

63 Kenneth Goodpaster, On Being Morally Considerable.

Cosmic Purpose Ethics

Cosmic purpose ethics may be described (roughly) in terms of their positions on the ultimate ends of evolution or the nature of God or God's purposes. They may take an approach which is theistic, non-theistic, pantheistic (God and nature are identical) or panentheistic (God is immanent in, but also transcends, nature) and generally can also be characterized in terms of whether they accept or deny the validity of the theory of evolution.

Fox makes three general points about cosmic purpose ethics. Firstly, in such ethics generally 'entities are seen as being intrinsically valuable to the extent that they are considered to embody or to be expressive of the ends (or values) that are under discussion...'[64] This results in there being a hierarchy of intrinsic values, with human intrinsic value commonly on top. This does not imply that those higher in any hierarchy can dominate those lower, but it does give a guide of sorts to resolution of conflicts between values despite it being frequently characterized in terms of *androcentric* (male-centred) reason. Secondly, Fox says that whatever the differences might be at a theoretical level between cosmic purpose ethics and other theories recognizing intrinsic value, at a practical level there is not much to differentiate them—entities in which intrinsic value is recognized should not be treated as mere means but, rather, should be treated as ends in themselves. Thirdly, cosmic purpose ethical theories may be more contentious than other intrinsic value theories primarily because many of the assumptions upon which cosmic purpose ethics rely are unfalsifiable and the theories are less open to modification than other intrinsic value theory approaches. This does not mean, however, that they are unable to be modified, although the likelihood of modification is lessened because there is no pressure to attempt falsification.

64 Ibid., p. 181.

The above assumes that ethical theories of environmental concern should be judged in terms of a narrow form of rationalism that excludes, for example, emotion from being taken into account. Deep ecologists and ecofeminists alike argue against such a narrow definition of rationalism.

Life-based Ethics

In his treatment of life-based ethics, Fox argues for a taxonomy that locates autopoietic activity as a subset of conative activity generally and that places 'biological ethics' and 'ecosphere (or Gaian) ethics' as species of autopoietic ethics. However, as shown later in this chapter, if there is to be something called autopoietic ethics, it must be disentangled from any ecosphere or Gaian ethics on the grounds that autopoiesis cannot validly be applied to all ecosystems. It is interesting to note that Fox makes autopoietic ethics a subset of conative ethics and that he uses the concept of autopoiesis and the concept of *conatus* in his exegesis of deep ecological views without acknowledging that the two are subsumed by the concept of *dūnamis*. As a result he fails to acknowledge that some of significant theories of environmental ethics rely explicitly or implicitly on morally-considerable entities having some form of immanent teleology.

Fox's claim that autopoietic activity is a subset of conative activity is interesting, but he does not explore the relationship fully. He bases his characterization of life-based ethics as *conative* on Arne Naess's treatment of self-realization which is derived directly from Spinoza's concept of *conatus*. But, the subsumption of autopoietic activity under conative activity is not at all clear for if Spinoza's notion of *conatus* is employed strictly, then *conatus* is immanent in all things and not just in living things, whereas autopoiesis claims to be a defining characteristic of *living* things. As shown in the previous chapter, there is a bifurcation in Spinoza's concept of *conatus*. All individual things, animate or inanimate, have the property of endeavouring to persist in their being—that is, they have a *conatus*.

All *living* things, however, have an additional *conatus* through which they *actively* pursue the immanent goal of their being, the goal of becoming a mature specimen of their species. There is a substantial equivalence between autopoiesis and *conatus* only so far as they apply to living things. But it is precisely living things that are the focus of this section, and more especially that *change* in living things that is definitive of living. This is something with which Aristotle was concerned in his *Physics*, notably that part centring on his concept of *dunamis*—and particularly the concept of an active *dunamis*.

With the aim of explaining, defining or otherwise determining what it is to be morally considerable, Aristotelian immanent teleology can be taken as a sort of umbrella concept to encompass the others. There is good justification for this, since autopoiesis attempts to provide an operational definition of living systems, one sense of *conatus* attempts to explain what it is that allows a living thing to persist in its being despite changes, and both of these can be expressed in Aristotelian terms. As will become clear, however, the appeal to Aristotelian teleology will not be sufficient to explain the role of ecosystems and the concept of supervenience discussed earlier will provide important help.

Generally, those who advocate a life-based ethics recognize that for an entity to be morally considerable it must *have* interests. As indicated earlier, morally significant interests are primarily welfare interests but preference interests also should be taken into account. An entity can be harmed through being inhibited or prevented from achieving, or even pursuing, its welfare and preference interests, and can be benefited by being permitted, or assisted, to achieve or pursue its welfare and preference interests. Life-based ethics hold generally that sentience is a sufficient condition for the possession of interests but that it is not a necessary condition: all, but not only, sentient entities *have* interests. Under a life-based ethics all living things have certain interests that need to be fulfilled if they are to continue living—i.e. 'welfare' interests. However, living things do not need to be 'aware' of these interests (not even in a very broad sense of awareness).[65] All biological organisms have interests and are—whether consciously or not—concerned with actively seeking some states of affairs rather than others. This might be taxic behaviour in the case of animals or tropic behaviour in the case of plants or the processes of homeostatic regulation.[66] From an Aristotelian position the state of affairs being sought by organisms is their own flourishing or *eudaimonia*.

To summarise briefly, the fact that life-based ethics may be expressed in terms of *conatus* has led to this position sometimes being characterized as *conative* ethics. This, however, is not necessarily a Spinozist view. In Spinozist terms, inanimate entities as well as animate entities strive to persist in their being i.e. all existing things have a basic *conatus*. Living things may have a qualitatively different *conatus* from that of non-living things, but non-living things have a *conatus* nonetheless.

Val Plumwood also represents a position that assigns moral value to inanimate as well as animate entities. She has a concern about limiting moral considerability to living individuals and suggests that there is a need to 'extend moral consideration to at least the wider class of beings with a *telos* of their own' and that 'drawing the moral boundary at living things has the problematic consequence that the wild river, the forbidding mountain and the venerable glaciated landscape on which the story of the earth's history and power is inscribed, have value only for and in virtue of the living things they contain or entertain.'[67] Plumwood's position is consistent with one

65 A useful parallel can be drawn here with Feinberg's distinction between 'harm' and 'hurt' and how the distinction hinges on knowledge. Someone or something can have their (its) interests harmed without knowing that this has happened but when the victim becomes aware of the 'harm' it is transformed into a 'hurt'. Feinberg, Social Philosophy, p. 27.

66 Fox, Toward a Transpersonal Ecology, p. 166.

67 Val Plumwood, Feminism and the Mastery of Nature, London: Routledge, 1993, p. 210.

that determines moral considerability on the basis of the possession of *conatus* (in its full sense). And given the relationship between *conatus* and *dunamis* established in the previous chapter, Plumwood's position implies that all things that have a *dūnamis*—active and/or passive—are morally considerable. Under such a position, because all things in the world have *conatus* in Spinoza's (full) sense, all things in the world would be morally considerable, even if they are not *equally* morally considerable. But the possession of only a passive *dūnamis* implies nothing about the possession of a 'good' or of 'interests' that can be promoted or inhibited—whereas possession of an active *dūnamis* does. For this reason, moral considerability should be admitted only for that class of entities that has an *active* potential (*dūnamis*). If entities with a merely passive potential do have value, it is not the same sort of value that living individuals have.

A problem with Fox's account is that he assumes that there is a third order of autopoiesis in general (i.e. systems that are unities) and that things such as ecosystems are living individuals whose status as individuals can be demonstrated using the concept of (third-order) autopoiesis. There is, however, considerable disagreement over whether there is a third order of autopoietic activity in general, and, even for those who acknowledge that a third order is possible, there is no agreement on the sorts of entities that might exhibit third-order autopoiesis.

Third-order Autopoiesis and Ecosystems

In working out his taxonomy of life-based ethics Fox at first suggests that life-based ethics might be divided into 1) biological ethics, where the prime entity considered to be morally considerable is a biological *organism*, and 2) autopoietic ethics, where what is morally considerable is defined in terms of autopoietic *systems*. But later he subsumes biological ethics under the class of autopoietic ethics on the grounds that autopoiesis as a criterion of moral considerability introduces a much larger class of morally considerable entities than that defined by the concept of biological organism. It may be the case that autopoietic ethics encompasses biological ethics, but, it is not the case that autopoietic ethics also encompasses an ethics in which all ecosystems are in themselves morally considerable.

As noted, a significant problem with Fox's claims about what an autopoietic ethics might encompass hinges on assumptions he makes about the systems nature of autopoiesis and the levels of complexity to which it can be applied. In making these assumptions he does not take account of the considerable controversy over just what sort of entities might validly be considered autopoietic. There has been, for example, a strong debate within systems theory over the applicability of autopoiesis to systems in general.[68]

Those who support an ecosystem or ecosphere ethic generally take the position that ecosystems (or the ecosphere) have an 'organismic nature' and can be seen as a whole and, as a 'whole', is taken to possess interests of an 'intrinsic kind. But, there are good reasons why ecosystems in general should not be taken as organismic wholes and why it cannot be simply assumed that the notion of autopoiesis can be extended beyond individuals where its application is already quite problematic.

Those, like Fox, who extend autopoiesis beyond individual organisms confuse the concept of a living individual with the concept of a living 'system'. In doing this they make a logical error. Living individuals might well be living systems, but as Brennan demonstrates as part of his general position that ecosystems can not be considered 'wholes' in the same way as individual organisms can, not all living systems should be considered individuals.[69] In all this it should be remembered that the term autopoiesis was originally coined to refer

68 See, for example, International Journal of General Systems, Vol. 21, 1992 – a special edition devoted to discussion of autopoiesis and that to which it might be applied.

69 Brennan, Thinking About Nature.

only to a process peculiar to individual cells and that there is considerable dispute over whether it may be applied to individual organisms, let alone beyond them. Confusion between the concepts of living individual and living system points to one of the problems at the heart of the concept of autopoiesis and at the heart of a biological ethic—the problem of what constitutes an *individual.* This can be examined in several ways, the most important of which is the difficult philosophical question of what an *individual* is. This will be taken up in the next chapter in the context of individualism in Paul Taylor's environmental ethics, the distinction that Mary Louise Gill makes between a horizontal and a vertical unity. Meanwhile, despite a question mark remaining over what individuals in general are, we can still fruitfully examine the ontological status of such things as ecosystems and come to the conclusion that ecosystems should be considered to have properties that supervene on the properties of biological individuals but are not themselves individuals.

As already noted, there is a considerable literature that argues against applying autopoiesis, beyond the level of the cell and individual organisms, to a so-called third order or third level that takes in groups, such as social groups and, by extension, ecosystems. If we were to impute moral value to ecosystems on the basis of the claim that they are autopoietic entities there would remain a substantial hole in the argument for the moral considerability of ecosystems. There may well be a value in ecosystems that makes them morally considerable entities, but autopoiesis can not be employed as a valid criterion for determining such value. The question of the moral status of ecosystems needs to be examined more closely.

For the purposes of discussion we can characterize a minimal configuration of so-called third-order autopoietic unities in terms of operational wholes comprised of aggregated individuals. On this basis, if there were a third level of autopoiesis it might take the form of a human social group such as a family, or a human social institution, such as a legal system, or even other non-human social groups such as whale pods, bird colonies or ant nests. Or, it could be that natural communities or ecosystems—on whatever scale, from the Earth conceived as a giant single ecosystem down to local and micro-level communities and ecosystems—meet the criteria to be considered third-order autopoietic unities. However, it should be remembered that Varela claims that to categorize families, legal systems etc. as autopoietic unities is to make a category mistake—it is to confuse autopoiesis with autonomy.[70] Despite this, ecosystems and communities present themselves, potentially at least, as special cases for moral considerability and are worth looking at more closely.

If possession of the property of autopoiesis is a sufficient criterion of moral considerability and if ecosystems or communities are autopoietic unities, then ecosystems or communities could be counted as morally considerable entities. But there are a number of problems about making such claims in respect of ecosystems or communities. One problem is whether one takes the basic unit to be a *community* or an *ecosystem*. A community includes only biotic elements, whereas an ecosystem includes abiotic elements in addition to biotic. It is rare, however, in environmental philosophy with a biocentric focus, for a clear distinction to be drawn between the concepts of community and ecosystem—there is usually a strong tendency for the term 'ecosystem' to be made to do the work of both. Giving 'ecosystem' such a dual role, however, impoverishes the discussion as it is less easy to distinguish precisely what it is that may be a source of value. Mathews, for example, looks at ecosystems as possible centres of value and talks of communities only in terms of *human* communities.[71]

70 Quoted by Vincent Kenny, 'On the Subject of Autopoiesis and its Boundaries,' p.187. An autonomous thing conforms to its own laws alone, it is independent. All autopoietic things are autonomous things, but the converse is not true. A weather system, for example, is autonomous but is not autopoietic. Autonomy is one the characteristics of autopoietic unities but is not sufficient to describe them. For the sufficient characteristics of autopoiesis, see my discussion in chapters 1 and 2.

71 Mathews, Ecological Self.

A key point to be addressed in any extension of autopoiesis to a third level—and for clarifying, in general, what an individual is—is the question of *boundary*. In the original formulations of autopoiesis, having a boundary *created and maintained by the entity* is a necessary property of an autopoietic unity. This point is emphasised by Varela in his *Principles of Biological Autonomy*, where he insists that what he calls a 'topological' boundary is necessary. Although it is not clear from Varela what a 'topological' boundary is, we can take it here to mean the presence of a *surface* which bounds the entity. Many things, though, have a boundary surface. What distinguishes an autopoietic unity—and biological individuals in general—is that it has a boundary surface that is created and actively maintained by the unity itself. It is the *boundary* that defines the system against its environment. But, an entity such as a social group does not have a boundary in the same sense as, say, a simple cell or in the sense that the skin is a defining boundary for humans and other animals. It may be true that social groups should not be considered individuals because they do not have a boundary surface, but it is not immediately obvious whether or not ecosystems or biotic communities have boundary surfaces and so (on the basis that they also meet the other criteria) it is not obvious whether or not they should be considered unities.

One outstanding candidate to be considered a third-order autopoietic unity is the *limit* case of terrestrial ecosystems: the planet Earth. The planet has a boundary in physical space, as testified by the images of our blue planet taken from the various space missions. As to whether this limit case of ecosystems is also an autopoietic system, consider the idea of the Earth-as-Gaia as suggested in Lovelock's Gaia hypothesis.[72] This hypothesis suggests that the biosphere is a self-regulating, self-maintaining system that might be conceived of as a

72 James E. Lovelock, Gaia: A New Look at Life on Earth, Oxford: Oxford University Press, 1979.

living unity. Fox[73] suggests that the planet-as-Gaia, that is, the biosphere, can be regarded as an autopoietic unity. The Earth-as-Gaia does seem to be a good candidate for an autopoietic unity—it is self-maintaining and has a self-produced topological boundary—but this says nothing about ecosystems in general also being unities. There may well be good reasons for people to consider ecosystems as unities: they are self-sustaining, self-repairing and there are clear stages of growth towards a climax community—the most mature instantiation of an ecosystem. But, it is not clear that they meet the other necessary criteria: they do not, for example, always have a distinct topological boundary and they *do* conform to Brennan's idea of an aggregate[74].

If ecosystems are unities they may be centres of value, but only if they also have a self-created, self-maintained boundary in physical space. None of the writers using the concept of autopoiesis to analyse environmental issues distinguishes—as Brennan does—between a unity and an aggregate that appears to function as a unity. For Brennan an ecosystem is not a whole in its own right, but is a set of properties that supervene on the properties of component parts. And if an ecosystem is not a true unity but is indeed only a set of properties supervening on the properties of certain aggregated biological individuals, then it cannot be an autopoietic unity. It is not immediately obvious, though, what the status of the whole Earth-as-Gaia is. It appears to meet the criteria for being a unity, but also can validly be conceived of as an aggregated whole. It is a key part of Fox's discussion that ecosystems are morally-considerable in their own right because they are autopoietic unities, so it is worth taking a closer look at the question whether ecosystems—or even communities—can indeed be autopoietic unities.

73 Fox, Toward a Transpersonal Ecology, p. 172. See also Elisabeth Sahtouris, Gaia: the Human Journey from Chaos to Cosmos, New York: Simon and Schuster, 1989, p.40

74 Brennan, Thinking about Nature.

There is no doubt that a savannah ecosystem (or even a savannah community), for example, is distinct from that of a neighbouring woodland even though the boundary is not well defined. The trees might thin out as one leaves the forest or become more numerous and closer together as one enters it, but there is definitely a point when one is either in the forest or in the savannah. There is definitely something that we first encounter when we 'bring forth' the system into our consciousness as a unity. The boundary between ecosystems can be mapped, even if imprecisely—but is this boundary topological, and if it is not, can an ecosystem be an autopoietic unity?

The question of whether ecosystems can be considered autopoietic unities can be tested using the original six key points set out by Varela, Maturana and Uribe.[75] Beginning with the notion of boundary, these points take the form of a questionnaire such that if there is a 'no' answer to any question then the system under consideration is not autopoietic. The points in question, then, are as follows:

1) Does the unity have identifiable boundaries? One example that demonstrates that ecosystems can have boundaries is an oasis. There is definitely a boundary to the oasis ecosystem marked by the presence or absence of water. However, although this is a mappable boundary, it is not immediately clear whether it is the same sort of boundary as that bounding an individual organism. For instance, is the boundary self-created and self-maintained? And is the boundary unique to the oasis ecosystem? Surely the boundary between the desert and the oasis is a shared boundary, in which case it is not clear whether one ecosystem or the other actually created it. If neither ecosystem can be said, unequivocally, to have created the boundary that distinguishes one from the other, can either be considered to be an autopoietic unity? The answer should be 'no', but let us look at the rest of the points before making a final judgment.

75 Varela, Maturana and Uribe, 'Autopoiesis: the organization of living systems, its characterization and a model.'

2) Are there constitutive elements or components to the boundary? Assuming (for the moment) that an ecosystem is a unity, it very obviously has components which inhabit and are constitutive of it—the trees, grasses, birds, insects etc.—and some of these will make up a boundary. The question, though, is whether what is perceived as a boundary is a boundary in autopoietic terms or whether it is just an aggregation of individuals the limit of whose spread is determined by an outside force. In the case of the desert and the oasis, the outside force might be as simple as the presence or absence of water. A similar argument can be made for the limit case of ecosystem, the Earth-as-Gaia.

3) Do the *components* of the unity themselves determine the interaction and transformation of the components? This can be understood in terms of the 'structure determinism' that is one of the characteristics of autopoietic unities. That is, each unity will only change in ways that are already allowed by the *structure* of the unity itself.[76] If the components are not determining their own transformations and interactions (or changes in general), then any changes are being brought about by external forces. To be an autopoietic unity an ecosystem must be such that any changes in it are determined by the ecosystem itself. This presents a possible problem because an ecosystem is at least a *process*, and anything that influences the process can be considered part of the process. On this basis, it would be difficult to conceive of an ecosystem being bounded at all. Although, if the concept of being bounded were to be something like the concept of 'localization,' then the problem would change. A process can be localized, even contained. For example an event is a process, but is also localized. A living organism can be considered a localized process. Part of the problem is that we have a commonly understood notion of organisms as things with clearly defined

76 For Maturana and Varela, structure is the 'actual relations which hold between the components', in a given space and at a given time: structure is the embodiment or the actualization of organization. Maturana and Varela, Autopoiesis and Cognition, p.138.

boundaries, so we can make the process of living something 'internal' to the organism, even though it might be difficult to specify the boundaries of the process. It is an open question as to whether the terms 'boundary' and 'localization' can be treated as synonymous in the context of autopoietic unities, but it should be noted that, in autopoietic terms, the creation and maintenance of a boundary is a continuous process undertaken by the unity itself.

4) Do the components of the boundary constitute the boundary through 'preferential neighbourhood relations and interactions between themselves'? That is, is the boundary determined by the properties of the components; is the boundary a property of the system itself and not imposed from outside; and is the boundary formed by neighbouring components 'preferring' one potential component over another? There seems to be a circularity operating with respect to this concept of preference. Preference between components appears to be one of the conditions for there to be a boundary, yet preference can be demonstrated simply by the presence of a boundary: if there were no preference between existing components there would be no self-created boundary. The mere fact that one component is contiguous to another can be seen as a demonstration of preference. Unless the boundary is to be determined entirely by external forces, the components of a boundary find neighbours which support them. And if the presence of an abiotic element such as water determines the boundary in some significant way, the boundary may be considered to be determined from outside. However, if the components are structurally determined to react in defined ways to the presence of water then the boundary can be considered to be internally determined. Here the distinction between community and ecosystem is important. In the case of a community, would water be an outside force, and, in the case of ecosystems, an internal force? This points to the probability that autopoiesis is not a fully-coherent concept and that its power of explanation is considerably less than that which has been claimed of it.

5) Are the components of the boundary produced by the interactions of the components of the unity? In the case of ecosystems,

the answer to this is yes and no. Yes, the components of the boundary of an ecosystem are produced by the components of the ecosystem, but, no, they are not produced *only* by the components of that ecosystem. At the second, organism, level of autopoiesis identifying a boundary is relatively straightforward: the components of the boundary of the organism are part of that organism and only that organism. At the level of an ecosystem, however, the components of the boundary also form the components of the boundary of an adjacent ecosystem. If the components of the boundary of ecosystem (or community) ‘A’ are also part of the boundary of ecosystem ‘B’ then the components of the boundary of ‘A’ are also at least partly produced by the components of ‘B’, and vice versa. So, the boundaries of ecosystems are not fully produced by the components of the ecosystem itself and so ecosystems should not be considered autopoietic unities.

6) Are the system components produced by the transformation of, and interaction between, components? In any ecosystem there are things entering and leaving that are an active part of the system’s process, but are not produced by the ecosystem. Take for example the migratory ducks in a wetland community: they fly in for a few weeks or months and then fly out again. They contribute to the system’s processes, but are not produced by them, or not entirely by them: the ducks are significant members of at least two geographically separated ecological communities (e.g. one northern hemisphere, one southern). The same could also be said of the gut flora that inhabit, say, the human digestive tract. They help digest food and so contribute to the well-being of their hosts. They may enter or leave the host, but they do not contribute to the autopoiesis of the host.

Summary

Although the application of some of the six points needs to be examined more closely, and despite some additional problems, It can be said with justification that ecosystems should not be considered (third-order) autopoietic unities and their alleged autopoietic status cannot be used to justify them being morally considerable. Ecosystems clearly have value, but not because they are unities.

Rather than be thought of as some sort of ontological unity, ecosystems might be better considered as aggregates, or as what Brennan would call non-unified part collections.[77] Ecosystems might be considered as unities for political purposes but this does not affect their ontological status. The habitat of the noisy scrub bird in south-western Australia is such a case. When the bird was rediscovered in the early 1960s, after having been thought extinct for most of the century, its habitat was declared a national park so that the species might be protected. Here its habitat and therefore the ecosystem of which it was part was determined arbitrarily by drawing lines on a map. The ecosphere (Earth-as-Gaia) though, as a limit case of ecosystems, does have a claim to be considered an ontological unity, although the claim is far from being substantiated. This indicates, in part, that autopoiesis is not a fully-coherent concept and does not explain why living things are centres of value or why ecosystems have value for themselves.

What we can say is that living things exhibit an active potential for change and initiate change from within. Each living thing, so defined, stands in a relation of subvenience to the ecosystem(s) with which it interacts. That is, the properties of anything that could be described as an ecosystem supervene on the properties of individual living things.

77 Brennan, Thinking About Nature, p. 80ff.

In short, living things are ontological unities in which an *active dūnamis* necessarily inheres. Neither ecosystems nor biotic communities can be considered to be ontological unities, but should be considered collections of properties that supervene on the properties of individual living things. In the sense that ecosystems are not singularities with an immanent *telos*, do not have an inherent active *dūnamis*, and are not autopoietic unities, they should not be considered to have interests that are uniquely their own—as individual organisms do. In Taylor's terms,[78] to be considered in the next chapter, ecosystems and communities do not have interests that can be benefited or harmed: any benefit or harm accrues in the first place to the individual (living) components of the system or community, although there may be a 'flow-on' benefit or harm to other members of the system through the whole being affected by a change in one or more parts.

It is clear that the idea of autopoietic ethics needs to be modified as a consequence of the fact that ecosystems and communities are not autopoietic unities, but the need for modification also derives from a certain incoherence in the concept of autopoiesis itself.

The next chapter examines Taylor's ethics of respect for nature—an ethics predicated on biological individuals having a unique inherent worth. The chapter asks how Taylor's ethic would fare if all and only individual living unities have an immanent *telos* and an internal *dūnamis* by which they initiate and direct changes in their own being.

78 Paul Taylor, Respect for Nature, Princeton: Princeton University Press, 1986.

4. 'Nature', Moral Considerability and Respect

Introduction

This chapter focuses primarily on the environmental ethic that Paul Taylor sets out in his book *Respect for Nature*. Taylor's is one of the most widely cited works on environmental ethics and is particularly interesting because of the key use it makes of the idea that organisms are teleological centres of activity, an idea that is implicitly Aristotelian—or which can be expressed in Aristotelian terms. His work warrants careful analysis and evaluation.

Taylor takes a biocentric (as opposed to an ecocentric) approach to environmental ethics in which individual living things are to be afforded the same respect, and in much the same way, as persons are to be afforded respect in a Kantian human-centred ethics. In Taylor's theory it is individual organisms that are *the* morally-considerable entities, and they are morally considerable because they are 'teleological centres of life' with a good of their own.

Taylor's theory has made a major contribution to environmental philosophy and there is much to learn from it, but there are problems in Taylor's approach, particularly when he relies on what he calls respect for 'nature,' but does not spell out sufficiently what he takes 'nature' to be. His theory can be supplemented and improved by addressing and clarifying the implicit Aristotelian conception of nature that is at its heart. Also, although Taylor wants individuals to be the primary morally considerable entities, his theory lacks a clear articulation of what a living *individual* is—of what a biological unity is. One account of the nature of an individual adequate to Taylor's purpose can be arrived at by analysing the paradox of unity in

Aristotle's treatment of substance. This can be done using Gill's detailed analysis on this point.[1]

Taylor's Biocentric Ethic

Taylor distinguishes between three fields of ethics: a human ethic, dealing with how human persons should act toward other human persons; a nature ethic, dealing with how moral agents should be guided in their interaction with what Taylor calls natural ecosystems and their wild communities of life; and an ethic of the bioculture that guides the conduct of humans in their treatment of animals and plants in artificial environments. He maintains that there is a symmetry between a human ethics based on respect for persons and a life-centred ethics based on respect for nature. The two systems of ethics are nevertheless distinct, with each having respect as an *ultimate* end. In the case of human ethics the ultimate end is respect for persons whereas for environmental ethics the ultimate end is respect for nature. Neither system can subsume or override the other, so that, for example, when a human's needs come into conflict with the needs of a living non-human the human's needs do not necessarily override the non-human's and vice versa. Of course, it is possible to have a nature ethic based on respect for persons by extending the definition of moral personhood. For example, moral personhood might be ascribed (typically) to sentient animals or even (less typically) to broader classes of things such as ecosystems or even the Earth itself (as Gaia) but that brings its own considerable problems.

Taylor deals with the ethics of the bioculture only very briefly. For him, such an ethic has two characteristics: a reliance on total

1 It should be noted that Stephen R.L. Clark has also put forward a similar interpretation.

human dominance and treatment of non-human living things as means to human ends. It is not clear, however, why different ethics should apply to 'wild' nature and to 'cultivated' nature, or even to humans and nature, since as Taylor himself argues all living organisms have a good of their own, and so their instrumental value (to humans) does not settle the question of how they should be treated. Although instrumental value is generally taken as the instrumental value *to humans*, there are other types of instrumental value: e.g. grass has instrumental value to an antelope and an antelope has instrumental value to a cheetah. Even domesticated animals and plants—which have instrumental value to humans—are teleological centres of activity and if being such is a sufficient criterion of moral considerability, as Taylor suggests it is, then they also should be covered by an ethic of respect for nature.

Taylor's environmental ethic concentrates on 'wild' nature, and adopts a biocentric viewpoint. That is, it recognizes that non-human living entities have a valid moral claim to have their interests considered and that situations where there is conflict between human and non-human interests require resolutions that take account of the competing moral claims. Once a biocentric position is adopted, humans can not validly argue that they have (*a priori* at least) a greater moral claim to survive than any other organism. For Taylor, adopting a biocentric point of view entails accepting both a conceptual and a normative claim. The conceptual claim is that animals and plants *can* meaningfully be thought of as *moral subjects* and that, as a result, plants and animals can be treated rightly or wrongly. The normative claim is that animals and plants *are* to be considered *moral subjects* and that we *ought* to treat them rightly and refrain from treating them wrongly.

Taylor maintains that there is a symmetry between an environmental ethic based on respect for nature and a human ethic derived from Kant's formula of the end-in-itself, and which focuses on personhood and the principle of respect for persons. His human ethic and his environmental ethic each have three main components: (1) acceptance of a belief system with respect to other moral agents;

(2) an attitude of respect—respect for persons in the case of the human ethic and respect for nature in the case of the environmental ethic; and (3) a system of rules and standards. This system of rules and standards allows one to prescribe what should be done by moral agents wishing to act morally. The attitude of respect mentioned under (2) is a moral attitude because it is, following Kant, universalizable and disinterested—that is, each moral subject is worthy of respect and, for Taylor, worthy of respect to the same degree. For Taylor all moral subjects thus have equal worth and the attitude of respect is not derivable from any other more general or fundamental attitude.

The first component of Taylor's environmental ethic, the belief system, involves a biocentric outlook on nature with three components. The first is that one should *identify* oneself as a member of the community of life on Earth. This, however, does not mean that those who identify themselves as members of Earth's community of life have to deny their personhood and so be denied consideration in terms of a human ethics. It is, for Taylor, simply a recognition of one's *biological* nature, a recognition that one is living. The second component is that each individual organism is 'seen as a teleological (goal-oriented) center of life, pursuing its own good in its own unique way'.[2] This need not be a conscious pursuit—any living thing, even a single-celled organism, tends to protect its own existence and promote its own well-being. The third component of a biocentric outlook is that one must reject the idea that humans are necessarily superior to other living things—biocentrism precludes a hierarchical view of nature. It follows from the biocentric view that there is an equality of worth between all moral subjects. Taylor maintains that if one adopts the biocentric outlook then the attitude of respect for the Earth's wild creatures is the only morally suitable attitude to adopt—although it is not clear why he limits the biocentric outlook to wild nature and does not take in all organisms, including domesticated ones. Taylor's pre-occupation with 'wild' nature gives rise to one of the limits to his

2 Ibid., p.44

theory. Why, for example, cannot a dairy cow or a sheep have inherent worth? Surely these have a good-of-their-own. So why cannot his theory be extended beyond 'wild nature'? He does not make this plain, but perhaps if the attitude of respect for nature were a dominant ethic then the husbanding of animals for human use (instrumental value) would be done differently, or at least greatly minimized.

Taylor's environmental ethic has a strong Kantian flavour in that it appeals to a universalized imperative. His human-centred ethic, too, is Kantian, but the environmental ethic differs in that the objects of ethical consideration are not limited to rational beings, but take in all (wild) living things. Under Taylor's environmental ethic, 'actions are right and character traits are morally good in virtue of their expressing or embodying a certain ultimate moral attitude [which is]...respect for nature'.[3] There are four dimensions to the attitude, the valuational, the conative, the practical and the affective. The valuational is the disposition to make judgments of value or worth, the primary disposition being to regard all wild living things in the Earth's ecosystem as having inherent worth. The conative is the disposition to have ends or to want to achieve goals. The general ends of an attitude of respect for nature are avoiding harm to wild things and preserving the existence of wild things. The practical is the disposition to act for certain reasons. It involves making judgements about alternative actions, deliberating on choices, deciding what to do and exercising the will to carry out the decision. The affective is the disposition to have certain feelings, especially that of feeling pleased about anything which tends to maintain wild communities.

Taylor's Kantian emphasis is clearly evident. For instance, he claims that 'people show genuine respect for nature only when they act or decline to act *out of consideration and concern for the good of wild living things*...such respect is not exhibited by actions or omissions that have the same effect on wild creatures but are not

3 Taylor, Respect for Nature, p. 80.

motivated in that way.'[4] That is, the motivation for action is of primary ethical importance. In the passage just quoted, Taylor continues the Kantian emphasis by saying that the end of any action must be thought of as an ethically obligatory end if the action is to be moral. Plumwood, while praising Taylor's aim to defend a biocentric ethic in which a person's 'true human self includes his or her biological nature',[5] criticises him precisely because he positions his theory firmly within a Kantian ethical framework—a framework that makes a strong division between reason and emotion and which keeps morality strictly within the domain of reason. This, says Plumwood, leaves no room for such virtues as 'care' and effectively de-legitimates any role for emotion in ethics. A role for care and emotion in ethics is discussed in more detail in Chapter 7.

Moral Considerability

One of the key questions to be addressed in any nature ethic is that of what counts as a morally-considerable entity. It may be that Goodpaster's notion that simply being alive is a necessary and sufficient criterion of moral considerability although there are reasons other than Goodpaster's for this. On the face of it there are strong parallels between the position argued for so far, Goodpaster's and Taylor's, in which what he calls 'teleological centres of life' have a good of their own and are therefore morally considerable. Taylor agrees with Goodpaster's view that the fact that all living things deserve moral consideration does not also mean that all living things have rights.

4 Ibid., p. 82.
5 Val Plumwood, Feminism and the Mastery of Nature, p. 166.

Let us take up first the question, of what should count as an object of respect or moral consideration. Goodpaster says

> neither rationality nor the capacity to experience pleasure and pain seem to me necessary (although they may be sufficient) conditions on moral considerability...Nothing short of the condition of *being alive* seems to me to be a plausible and non-arbitrary criterion. What is more, this criterion, if taken seriously, could admit of application to entities and systems of entities heretofore unimagined as claimants on our moral attention (such as the biosystem itself)...our [current] paradigms of moral considerability are individual persons and their joys and sorrows. I want to venture the belief that the universe of moral consideration is more complex than these paradigms allow.[6]

This passage from Goodpaster is important for what it does not say as well as for what it does. He does not say that the biosystem *is* a morally considerable entity, only that it *may* be a claimant on our moral attention. He does not suggest that subsystems of the biosphere might be morally considerable, but refers only to *the* biosystem, presumably in the sense of Gaia that Lovelock later spelled out. And he does not address the ontological status of the biosystem. That is he does not state whether it should be considered as an *individual* (what Brennan would call a unified collection of parts), or as a mere aggregate of parts. What is perhaps most significant in this passage, however, is Goodpaster's rejection of the paradigm that views moral considerability in terms of *individual persons*—although this is not to deny that there are many ways to understand what a 'person' is.

Elsewhere in his article, Goodpaster claims that being alive is the only necessary criterion of moral considerability. In the course of arguing for this claim he puts forward what he sees as the main objections to it and then attempts to rebut them. Of particular significance is his third objection, which states that 'consideration of life can serve as a criterion only to the degree that life itself can be given a precise definition; and it can't.' His response to this is in part

6 Kenneth E. Goodpaster, 'On Being Morally Considerable,' p. 51.

to question whether a 'criterion of moral considerability must be strictly decidable in order to be tenable'.[7] He does, however, attempt a definition of life that has some precision—he cites Sayre, who claims that:

> The typifying mark of a living system...appears to be its persistent state of low entropy, sustained by metabolic processes for accumulating energy, and maintained in equilibrium with its environment by homeostatic feedback processes.[8]

This, says Goodpaster, may need qualifying but it strikes him as plausible in its own right and ethically illuminating because it suggests that at the core of moral concern is a respect for self-sustaining organization and integration—despite pressure toward entropy. Goodpaster could just as well be advocating an autopoietic ethics or an ethic of respect for nature in which the concept of 'nature' is informed by concepts from cybernetics and theoretical biology.

Individual 'Living Things'

Taylor agrees with Goodpaster's claim that all *living things* are morally considerable, but since the notion of what a living thing is, is as Goodpaster notes, not unproblematic, the sense in which Taylor agrees with Goodpaster remains to be clarified. A working definition is clearly needed of what it means to be a 'living thing', but Taylor employs no such definition and has been accused by Fox of using a biologically unsophisticated notion of what it is to be an organism.[9] This lack of sophistication, if proven, is significant because Taylor argues for a *biocentric* approach in which (following Goodpaster)

7 Ibid., p. 62.
8 K.M. Sayre, Cybernetics and the Philosophy of Mind, NY: Humanities, 1976, p. 91. Cited by Goodpaster, 'On Being Morally Considerable', p. 62.
9 Fox, Toward a Transpersonal Ecology, p. 174.

individual organisms, and only they, are morally considerable entities. Hence in Taylor's view ecosystems are not seen as being morally considerable in their own right, and as a result he does not account adequately for the moral considerability (or otherwise) of ecosystems. A theory based on biological individuals being morally considerable needs to have a clear account of what these individuals are, and Taylor's theory may well be flawed if it turns out to be predicated on a largely unargued and possibly simplistic assumption of what an organism is.

Taylor defines a *living thing* (i.e. individual) as 'a unified system of organized activity, the constant tendency of which is to preserve its existence by protecting and promoting its well-being'.[10] Now, while this definition is open to criticism, it is not as unsophisticated as Fox makes out. It is, for example, not very different from Sayre's cybernetic account of a living system cited earlier and that Goodpaster finds 'empirically respectable.' There are two noteworthy points in this definition: *living things* have a *conatus* in that they strive to persist in their being (or in his words: living things have a constant tendency to preserve their own existence); and living things also have a 'good of their own'. The idea of things having a good of their own is discussed below, where it will become clear that there is a strong Aristotelian flavour to Taylor's use of this. These two points, along with the discussion of the previous paragraph, combine to show that Taylor's definition of a living organism is consistent with recent definitions from theoretical biology—and, *prima facie* at least, is implicitly Aristotelian.

But why is it important to try to assimilate Taylor's theory (amongst others) to an Aristotelian position? We can begin to answer this by noting that if an environmental ethic needs to take into account the idea that all living things are morally considerable then it needs also to account for the nature of living things themselves—and

10 Taylor, Respect for Nature, p. 45.

accommodate their differences. Such an account of the nature of living things would not focus on the peculiarities of this or that individual or species but would look to find the essential property or properties common to all living things and would have to provide reasons why those things that possess this property or properties are morally considerable. In their own ways the theory of autopoiesis and Aristotle's teleological biology each tries to identify the essential properties of living things and—in recognizing that each living individual has a good of its own—gives at least the ground for moral considerability. As already seen, each living thing has a unique value simply because it exists—the value that comes from having an immanent *telos*. Taylor rests his ethic of respect for nature on the unique value possessed by each individual organism, so it will pay to look more closely at it.

A Kantian Emphasis

Taylor wants to extend moral considerability to all individual living things—which then become the subjects to be considered in a modified Kantian ethic. He replaces the *respect for persons*, a central component of a Kantian human-centred ethic, with the notion of *respect for nature*. In setting out his moral theory, he makes a distinction between the *formal* and the *material* conditions needed for valid moral principles, with the *formal* conditions defining what is meant by describing or classifying a rule, standard or principle as a moral one. Taylor lists five formal conditions for any set of rules and standards to constitute a valid normative ethical system. The conditions are, in a clearly Kantian move, that the rules and standards must be: general in form; universally applicable to all moral agents; intended to be applied disinterestedly; advocated as normative principles for all to adopt; taken as overriding all non-moral norms. Taylor's *material* condition for a human ethic, again following Kant, is *respect for persons*. The *material* condition for Taylor's environmental ethics is *respect for nature*. However, Kant's apparent insistence on the primacy of the value of human persons might seem to imply at first glance that one cannot legitimately extend Kantian ethics to all the individual things of nature. But, as shown later, it would be wrong to assume that Kant had a view of moral considerability characterized by extreme human chauvinism. This may also be true of John Stuart Mill, contrary to common readings, in that both he and Kant were prepared to admit more than only humans into the class of morally considerable entities.

In Kant's ethics, respect for persons can be seen in the version of the categorical imperative known as the 'formula of the end-in-itself' which demands that we treat 'humanity in your own person or in the person of any other never simply as a means but always at the same time as an end'.[11] But Kant did not limit his realm of moral considerability to humans alone. The main condition he placed on the notion of moral considerability was that to be morally considerable a creature had to be *rational* and the formula of the end-in-itself was to be applied to 'man *and in general every rational being*'.[12] John Stuart Mill also saw the realm of moral considerability extending beyond the human. For example, Mill said: 'the (Utilitarian) standard of morality...may...be defined as the rules and precepts of human conduct, by the observance of which an existence such as has been described might be, to the greatest extent possible, secured to all mankind; and not to them only, but, so far as the nature of things admits, *to the whole sentient creation*'.[13] The class of morally considerable entities is thus expanded by Kant to include all rational beings and by Mill to include all sentient beings: sentience becoming the sufficient criterion of moral considerability. If the class of morally-considerable entities can be expanded to take in all sentient

11 Immanuel Kant, Groundwork of the Metaphysic of Morals, p. 95.

12 Ibid., p. 95, italics added.

13 John Stuart Mill, Utilitarianism, Alan Ryan ed. John Stuart Mill and Jeremy Bentham: Utilitarianism and Other Essays, Harmondsworth: Penguin, 1987, p.283. Italics added.

beings, why can it not be expanded further? Taylor does take it further.

But the question of what is to be morally considerable is just one part of the issue. An ethic needs both to determine what entities are morally significant and to specify how to behave toward them. That is, it needs to set up what counts as moral *subjects*—sometimes also called moral 'patients'—and to specify in what ways the actions of any moral *agents* are to be constrained when those actions are directed toward a moral subject. Put another way, for an ethic to be put into practice there needs to be some entity that *considers* others, because 'other-regardingness' is an essential component of any ethic, and there needs to be some entity that is morally *considerable*.

For Taylor, a moral agent is 'any being that possesses those capacities by virtue of which it can act morally or immorally, can have duties and responsibilities, and can be held accountable for what it does'.[14] A moral subject is an entity toward which moral agents have obligations. That is, a moral subject possesses certain qualities or properties that determine in some degree how a moral agent should act toward it. All moral agents are also moral subjects, but not all moral subjects are moral agents.

An important element in Taylor's theory is that it is always possible for a moral agent to take a moral subject's standpoint and on the basis of this make judgements about how the moral subject ought to be treated. Arne Naess makes the same point with his notion of 'identification': we (human persons) feel empathy for non-human living entities because we can identify with them, that is, we can put ourselves in their position.[15] The ability of a moral agent to take an individual organism's standpoint is, however, itself dependent on moral agents conceiving those individual organisms as what Taylor

14 Taylor, Respect for Nature, p.14.

15 Arne Naess, 'Self-realization: An Ecological Approach to Being in the World', John Seed et al. eds., Thinking Like a Mountain: towards a council of all beings, Philadelphia/Santa Cruz: New Society Publishers, 1988.

calls 'teleological centers of life'[16] or activity, a notion that is central to Taylor's theory. Teleological centres of activity have a 'good' of their own, the possession of which enables a moral agent to recognize that the entity with such a good has inherent worth and is entitled to moral considerability. The key terms here are, 'teleological center of activity', 'good of its own' and 'inherent worth'. These need to be examined further.

Living Things as (Aristotelian) Teleological Centres of Activity

Taylor's explanation of how individual organisms are teleological centres of life is undertaken in Section 3 Part 4 of *Respect for Nature.* Although Taylor does not cite Aristotle here—in fact Aristotle is mentioned only in two footnotes—his notion matches in all important respects the Aristotelian notion of internal or immanent teleology discussed in Chapter 1.

In order to show that Taylor's notion of organisms as teleological centres of activity[17] is substantially Aristotelian, two main points need to be clarified: what is Taylor's conception of 'nature'; and what is an 'individual' organism. For reasons that will become clear, however, it is not possible to treat these two issues in isolation. Clarification of them will come down to a clarification of the difference between two forms of nature: the whole of nature and the 'nature' or essence of those individual components that together comprise the whole of

16 Taylor, Respect for Nature, p.125.

17 The phrase 'teleological centre of activity' is preferable to the phrase 'teleological centre of life' because it avoids begging the question when used as a definition of life.

nature. To begin with, let us see how Taylor puts his position. He says,

> we conceive of the organism as a teleological center of life, striving to preserve itself and realize its good in its own unique way. To say it is a teleological center of life is to say that its internal functioning as well as its external activities are all goal-oriented, having the constant tendency to maintain the organism's existence through time and to enable it successfully to perform those biological operations whereby it reproduces its kind and continually adapts to changing environmental events and conditions. It is the coherence and unity of these functions of an organism, all directed toward the realization of its good, that make it one teleological center of *activity*...All organisms, whether conscious or not, are teleological centers of life in the sense that each is a unified, coherently ordered system of goal-oriented activities that has a constant tendency to protect and maintain the organism's existence...'[18]

There are in this passage key phrases that point to a biological teleology which is fundamentally Aristotelian—phrases such as 'striving to preserve itself', 'realization of its good' and 'internal functioning' tending 'to maintain existence'. As demonstrated in Chapter 2, the idea that organisms strive to preserve themselves can readily be expressed in terms of the concept of *dūnamis*, an essential element in Aristotle's immanent teleology. Implicit in the idea that an organism's internal functioning and external activities are goal-oriented are the two forms of *telos* (internal and external *telos*), discussed in Chapter 1, that Aristotle uses to explain change or movement. And in order for these two forms of *telos* to operate in an entity, the entity needs to have an active *dūnamis* operating on itself and on things external to itself.

The idea that the organism's internal functioning has the constant tendency to maintain its existence through time is an especially important one for the individualism at the heart of Taylor's ethic of respect for nature. This individualism presumes that there is a solution to the problem of how an entity can be one thing at *this* time and still

18 Taylor, Respect for Nature, pp. 121-122. Italics added.

be the same thing at *another* time—what Gill calls, as noted above, the paradox of unity[19]—and something explored in more detail below.

This passage from Taylor also contains the idea that organisms are teleological centres of *activity*. That is, the idea that each organism can initiate action or movement in another and in itself. The idea that each living thing has within it the source of its own end-directed movement is central to Aristotle's biological teleology, as was pointed out in Chapter 1. There, the concept of internal end-directedness was associated with Aristotle's concept of *dūnamis* and its close correlates *energeia* and *entelecheia*. From the passage quoted above it is evident that—because the concept of a teleological centre of activity can be traced to Aristotle—Taylor's conception of an individual organism relies on notions that are implicitly Aristotelian. It suggests also that Taylor's conception of *nature*—that which he says moral agents ought to respect—might have a basis in Aristotelian concepts. But what Taylor means by nature needs clarification.

Taylor's 'Nature' and Teleological Centres of Activity

When Taylor says that the fundamental principle of his environmental ethic is 'respect for nature' he is apparently referring both to nature as a whole and to each of the individual components of which nature is comprised. (For the sake of brevity in this discussion of Taylor, the whole of nature will be expressed as Nature—with a capital—and individual nature will be lower case.) He does not make the distinction between the two 'natures' explicit, although he needs to if he is to maintain consistently that 'individual' organisms are teleological centres of life or activity. Taylor sees the value of Nature in terms only of the sum of the values of (living) individuals. But while, as was seen in the previous chapter, there may be no value in Nature in its

19 Mary Louise Gill, Aristotle on Substance.

own right,[20] Taylor needs to explain more clearly the distinction between the whole of Nature and its individual components.

To clarify the distinction between 'Nature' and 'nature', and to get a sense of just what it is that we should respect, it is worthwhile referring back to Chapter 1 and Mourelatos' suggestion that Aristotle made use of the concept of *phūsis* (nature) at two levels: as the sum of all essential powers *and* as an innate urge to persist in being. The 'powers' to which Mourelatos refers are given in terms of Aristotle's *dunameis,* of which there are also two levels—a general level and a smaller second level. According to Mourelatos' account the smaller group of powers constituting the essence is an entity's nature or *phūsis.* Within this set of essential powers is also an intrinsic power that involves no reference to specific other things. This intrinsic power is a special sense of *phūsis*, a 'principle of motion and rest...an innate urge for change [which is a] reflexive power, a power of self-development'.[21]

Manufactured items have the second level of powers accidentally whereas natural products have this level essentially. Mourelatos uses the shorthand PHŪSIS (capitalized) for this second sense of a '*phūsis'* which is intrinsic to—is the essence of—a thing. Both Mourelatos and Gill[22] argue that when *phūsis* is defined in terms of an innate urge (PHŪSIS), it becomes a species of the genus *dūnamis*. Taking the argument in Chapter 2 into account, this relates the second sense of *phūsis* very strongly to the notion of *conatus*, the role of which in relation to life-based environmental ethics is discussed in the next chapter. And, as noted in Chapter 1, Gill argues that there is an active

20 The argument that Nature as a whole, in the form of the concept of the biosphere, does not have value in and of itself is also made by Robin Attfield, The Ethics of Environmental Concern, Oxford: Basil Blackwell, 1983, p. 159.

21 Mourelatos, 'Aristotle's 'Powers' and Modern Empiricism,' pp. 99-100. Note that Mourelatos actually uses the terms "PHYSIS" and "physis". For the sake of consistency and conformity to contemporary practice, these are represented as "PHŪSIS" and "phūsis".

22 Gill, Aristotle on Substance.

motive principle (*dūnamis*) within a living thing that causes change in the thing itself. This active motive principle is the entity's *nature* (*phūsis*), a 'source of change in the thing itself qua itself'.[23] This active motive principle in a living thing is, in one sense, its *nature*. It is this inner motive principle, this species of *dūnamis*, this PHŪSIS, that makes a living thing a teleological centre of activity. And it is this inner motive principle that makes a living thing intrinsically valuable. If we are to show respect for nature, as Taylor argues that we should, it is in no small part this specific nature, this species of *dūnamis*, this PHŪSIS, to which respect is due—for it is this that sets the living apart from the non-living. It is the possession of this PHŪSIS that makes an entity morally considerable, as all and only those things that possess this PHŪSIS have a good of their own and are ends in themselves. The 'nature' that one ought to respect in something like Taylor's ethic of respect for nature is the PHŪSIS, or inner motive principle, as this is the source of value in Nature as a whole.

Addressing the Paradox of Individual 'Unity'

As suggested earlier, one of the key problems in Taylor's theory is that he does not make sufficiently clear what he means by an individual, especially in relation to the biological individuals that he urges us, as moral agents, to respect. The previous chapter has a partial solution to the problem of individuality in the discussion of Brennan's distinction between a whole that is a sum of parts and a whole that is a unified part collection. There a biological individual is shown to be categorically different from an ecosystem. The concern now is to identify more clearly what an individual *per se* might be, since the provision of such a definition is a problem for Taylor, for the

23 Ibid., p.218.

theory of autopoiesis and also for Aristotle. Here an especially useful argument set out by Gill provides valuable help.

Gill argues that 'organic composites, such as human beings and horses, are both ontologically and conceptually primary', where a unity is conceptually primary 'if it is a definable unity, that is, if it is definable simply in terms of itself and without reference to entities that are prior to it.'[24] In this her position is substantially the same as Brennan's. She differs from Brennan, however, in describing what she calls the 'paradox' of unity. Brennan sees an individual as a unified part collection but does not recognize any inherent paradox in the concept of unity. Gill describes the 'paradox' of unity by first noting that Aristotle believed all changes involve something that persists through the change, a continuant. She then distinguishes between *horizontal unity*, which is the requirement for unity through change, and *vertical unity*, which is the requirement for a definable unity. Gill says 'an entity is a vertical unity only if its nature can be specified without mentioning one thing in another, where the two items mentioned are distinct from each other...Vertical unity guarantees an entity conceptual priority because a vertical unity is not defined by something else that is prior to it'[25]. Aristotle argues in *Metaphysics* Z that substantial forms are, in Gill's terminology, vertical unities. Horizontal unity, on the other hand, is necessary for something to change but stay the same. For example, over the next month most if not all of the individual molecules in your body will be exchanged for different molecules of the same kind, yet you will remain in nearly all respects the same individual. Horizontal unity is necessary for this to happen. For Gill, the paradox can be put as follows:

> How can the requirement for unity of the continuant over time be reconciled with the requirement for unity of the whole generated composite at a time?...the

24 Ibid., p. 4.
25 Ibid., p. 145

> kind of unity needed to account for change conflicts with the vertical unity required of those entities that are primary substance.[26]

Proponents of autopoiesis maintain that the theory of autopoiesis can explain how a unity can change over time yet remain the same entity. In Glanville's terminology, autopoiesis claims to explain how something can be 'the same, but different'.[27] The concept of 'unity,' however, is still problematic in the literature on autopoiesis. Brennan's discussion does not go to the heart of the matter either. Brennan provides a good working distinction between organism and ecosystem, but even he admits that his account does not clarify sufficiently what constitutes the nature of a unitary object. He does suggest, however, that it is intuitively clear how a genuinely unified thing is distinguishable from an artificial entity. Brennan's view on wholes is substantially an Aristotelian view that makes use of the notion of causal powers in a way similar to that in which Mourelatos uses it. Taylor does not deal with the problem of change and identity in individual organisms and his theory can benefit from clarification of the problem of unity.

Some commentators argue, on the basis of passages in Aristotle's *Metaphysics* Z and H, that Aristotle thought generated composites (e.g. organisms) did not have what Gill calls vertical unity and that forms were conceptually primary entities.[28] But Gill claims that *Metaphysics* H.6 and Θ indicate a different conclusion, and in conjunction with parts of *De Anima* clearly suggest that 'composites of matter and form are unities...in some strict sense'. These passages also indicate that the doctrine of potentiality (*dūnamis*) and actuality

26 Ibid., pp. 6-7.

27 Ranulph Glanville, 'The Same is Different,' Milan Zeleny ed., Autopoiesis: A Theory of Living Organization.

28 For a general introductory discussion of this that makes sympathetic use of Gill's argument, see Hugh Lawson-Tancred, 'Aristotle', A.C. Grayling ed., Philosophy: A Guide Through the Subject, Oxford: Oxford UP, 1995, pp. 398-440.

(*energeia/entelecheia*) is crucial to solving the problem of unity for material composites.

Let us consider the paradox of unity more closely. If a composite (in the present case, an organism) is to be a primary substance in Aristotle's sense, a fundamental or basic unity not reducible to component parts, then it has to have vertical unity, that is, it has to be a definable unity. But composites are generated from form and matter. And generation, because it takes time, requires horizontal unity. So how can composites be vertical unities and horizontal unities? Aristotle's approach is to insist 'that there is a continuant which guarantees that the generation of something new is not sheer emergence of something out of nothing'.[29] That is, there is some part of the thing that remains the same before and after any change occurs. However, if the continuant is a distinct subject within the composite, the composite itself cannot have vertical unity. This comes about because anything whose nature is specified as one thing in another cannot have the property of vertical unity. Aristotle further claims that the continuant need not be *actually* present, but could be *potentially* present. He uses an account of chemical combination to make his point. Gill summarizes Aristotle's argument as follows,

> The ingredients of a compound (*mixis*) exist separately as those ingredients before they enter into the combination and will exist separately again once the combination has been dissolved. While the compound exists, the ingredients, although not destroyed, are not actually present either—their potentiality alone is somehow preserved.[30]

Sodium atoms, for example, combine with chlorine atoms to form common salt (NaCl). The compound NaCl is a stable compound.

29 Gill, Aristotle on Substance, p. 145.

30 Ibid., p.148. Gill notes that on this point Joachim (1922) and Bogaard (1979) give interpretations of Aristotle's theory which are different from hers and from each other's. Joachim says the ingredients of a compound are preserved 'at a lower grade of being'. Bogaard says the ingredients are 'screened from our view' because they are modified by new powers gained.

While the compound exists it can be argued that the components are not actually present because the compound does not exhibit the qualities or properties of the components. Pure chlorine is a highly-reactive elemental gas; sodium is a metal. The sodium and the chlorine can be extracted from the salt—but only by destruction of the salt. The salt retains the potential to produce chlorine and sodium but the sodium and the chlorine are in at least one sense not actually present, because the compound does not demonstrate the properties of the components.

In biochemical terms an organism is an organized system of sustained chemical reactions. While the organization holds things together, so to speak, the component elements do not have an independent existence, whereas the organism does. But when the organization begins to break down—i.e. when the organism begins to die—the components are free to take on their original form, to exhibit their essential properties.

Following Gill's clarification of the paradox of unity in Aristotle, in an organism, the ingredients lose their individual identity while they are part of the unity. The potentiality of the ingredients is preserved, but they are prevented from becoming actual again (which would lead to the disintegration of the organism) by an internal organizing principle. If the organism reverts to a mere collection of its component parts it ceases to be a unity and ceases to be a living thing. In fact, the point at which an organism begins to break down into its component parts may be a defining moment in the death of the organism. The internal organizing principle (which in Aristotelian terms is a *dunamei,* a power) allows actual ingredients (e.g. molecules) to be replaced by other actual ingredients while also allowing the unity to maintain its identity. The important concept of *organization* in autopoiesis can be explained in Aristotelian terms as that which helps components maintain their *potential* (to persist as what they are) while they are part of the *actual* structure of the existing entity.

So, what are genuine, primary, substances for Aristotle? What may count as a unity? Aristotle, says Gill, rejects both the elements and the parts of animals [as candidates for being a unity] but leaves room for whole plants and animals. He excludes the elements, saying 'none of these [earth, air, fire, water] is one, but like a heap, until it is worked up (*pephthē*)[lit. 'concocted'] and some unity comes to be from them' (1040b8-10). This statement indicates both that the elements fail to be substances because they are more like heaps than real unities and that certain entities generated out of the elements are unities and therefore substances. In the same passage he excludes the parts of animals and gives his reason that, if they are removed from the whole animal, they lose their ability to function as the parts that they were. Since he says nothing against whole plants and animals, these composites apparently escape the purge to take their place among (and perhaps pre-eminently among) the entities regarded as 'unities' that come to be from the elements.'[31]

Just as Brennan distinguishes between a mere sum of parts and unified, causally-integrated wholes with unique powers, the position outlined by Gill should be distinguished from a merely descriptive conception of unity, such as might be employed in describing an ecosystem as a unity. An ecosystem might be regarded as a unity for the purposes of argument, particularly a political argument, but it cannot be regarded as a *basic* unity in the same way that an organism should be so considered. In a basic (Aristotelian) unity the ingredients lose their *actual* separate existence (while keeping it potentially). On Gill's interpretation of Aristotle, organisms are such basic unities. This squares also with Brennan's conceptualization of wholes—and if we add to this his use of supervenience we arrive at a useful distinction between parts and wholes, and between organisms and ecosystems. As 'ingredients' of an ecosystem, organisms manage to maintain their actual separate existence all the time they are part of the larger whole. The ecosystem can, on the other hand (following Brennan), be seen as the sum of the essential properties of the components plus the sum of the supervenient properties each component has as a result of coming together with others. To call an ecosystem a unity in the same sense that an organism is called a unity

31 Gill, Aristotle on Substance, p.112.

is to make a category mistake. As Brennan has shown, although an organism such as a tree is a combination of parts, it is not the same sort of combination as is, say, the tree-plus-compost heap in the corner of his garden. It is clear that by using either Gill's argument or Brennan's it can be shown that an ecosystem is an aggregate of parts and is not a unified whole. An ecosystem does not have its own *dūnamis*—does not have powers in its own right—whereas an organism does. The advantage of Brennan's position is that it offers an explanation of the relationship between an organism and its environment i.e. a relationship of supervenience.

An ecosystem describes the interaction of organisms and in so doing describes some at least of the functional (supervenient) properties of the aggregated organisms which comprise it. Organisms on the other hand are categorically different. (They are simple wholes of a natural kind, whereas ecosystems are collections of parts). Organisms, as teleological centres of activity, are in a very important sense self-defining: their *telos* is determined internal to themselves. It is true that they have ecosystemic (functional) properties, but these are supervenient on properties that are basic or essential. An ecosystem does not have basic or essential properties and cannot therefore be considered a unity or an individual.

Inherent Worth and 'Good'

To understand Taylor's environmental ethic of respect for nature requires us to understand two key concepts: the idea of the *good* of a being, which Taylor says derives from the concept of the good in Plato and Aristotle,[32] and the (logically independent) idea of the *inherent worth* of a being. Having a good of its own is a necessary

32 Taylor, Respect for Nature, p. 61n.

condition for an entity to have what Taylor calls *inherent worth*, which he distinguishes from both intrinsic value and inherent value. Inherent worth is, for Taylor, the 'fundamental value-presupposition of the attitude of respect'.[33] Although having a good and having inherent worth are logically independent, it is necessary that a moral agent recognizes both if he or she is to adopt the 'attitude of respect' that Taylor says a moral agent should. A moral agent who adopts an attitude of respect for nature recognizes that an entity worthy of respect has both a good of its own and inherent worth.

But what are the entities that have a good of their own, have inherent worth and are therefore to be held as objects of respect? In Taylor's theory all animals and all plants, as individuals, have a good of their own which they strive to protect or promote. However, this use of the standard division between plants and animals is indeed one place where Taylor might be said to show a lack of biological sophistication. The old binary division between plant and animal kingdoms is now generally replaced by five categories: monera, protoctista, fungi, plantae and animalia —although the classificatory system in recent years has been in flux, with a number of claims made for significant change. (Between the five kingdoms or categories there is a primary division between the prokaryotes and the eukaryotes. The only prokaryotes - entities with no nuclear membrane - are the monera, or bacteria. The other four categories are constituted by various forms of eukaryotes —entities possessing nuclear membranes. Of these, the protoctista are defined as those eukaryotes that are not fungi, plants or animals.) Given such a revised biological taxonomy and the fact that the taxonomy itself is in flux, it may be more appropriate to rephrase Taylor and say that all *living things* have a good of their own, where living things can also be described as entities having a good of their own and an internally-determined directiveness. This, of course, is a more strongly Aristotelian position.

33 Ibid., p. 71.

Any entity with a good of its own has what Taylor calls *inherent worth* and is therefore worthy of respect. The concept of inherent worth will be examined shortly, but for now it is enough to point out that it follows that, if a moral agent recognizes inherent worth in something, then the moral agent will adopt an attitude of respect toward the thing. This is in line with Taylor's predominantly Kantian approach to his ethics under which moral agents have a duty to respect that which has a good of its own—because it is an end-in-itself.

Taylor contrasts his use of inherent worth with the related, but significantly different, notions of intrinsic value and inherent value. The terms, although of crucial importance, are defined in various ways. For example, Holmes Rolston says intrinsic value is 'value as an end in itself without necessary contributory reference'[34] whereas Freya Mathews' concept of intrinsic value is that it is the value a self assigns to itself on account of its status as a self, regardless of whether the self is valued by external valuers. Intrinsic value for her is a 'fact' of the world[35] whereas inherent value 'is the value that things have on account of being valued for themselves—rather than for their utility—by external valuers'. She notes that Callicott affirms both these non-instrumental notions of value but sees the notion of inherent value sufficing to do the work of both.[36] Mathews agrees that Taylor's use of inherent worth is closely related to her use of intrinsic value. In Taylor's scheme, any activity has *intrinsic value* for those who find satisfaction in it if the activity is 'carried on for its own sake or as an end in itself, whether or not it is also valued for its consequences' whereas the *inherent value* of anything is 'relative to and dependent upon someone's valuing it'. *Inherent worth*, on the other hand 'is to be

34 Holmes Rolston, III, 'Duties to Ecosystems,' J.Baird Callicott ed., Companion to 'A Sand County Almanac', Madison: University of Wisconsin Press, 1987, p. 268.

35 Mathews, The Ecological Self, p.105 and pp. 117-130. In pp. 178-179 Mathews also provides a good summary of some of the main definitions of intrinsic value and cognate terms.

36 Ibid., p. 178.

attributed only to entities that have a good of their own,' and an entity with inherent worth has it 'regardless of any instrumental or inherent *value* it may have and without reference to the good of another being'.[37] To say an entity X has inherent worth in Taylor's sense is to assert that

> a state of affairs in which the good of X is realized is better than an otherwise similar state of affairs in which it is not realized (or not realized to the same degree), (a) independently of X's being valued, either intrinsically or instrumentally, by some human valuer, and (b) independently of X's being in fact useful in furthering the realization of some other being's good, human or nonhuman, conscious or nonconscious.'[38]

Having this *inherent worth* does not imply or entail having moral *rights,*[39] but those entities that have inherent worth are, in Taylor's scheme, morally considerable—that is, right or wrong can be done to them and they ought not be treated as mere means to human ends. This principle of not treating morally considerable entities as mere means to an end is an extension of Kant's expression of the categorical imperative in the principle of the end in itself.

37 Taylor, Respect for Nature, p.75.

38 Ibid., p. 75. Taylor acknowledges (p. 75n.) that his concept of inherent worth is 'essentially identical' to the concept of inherent value which Tom Regan develops in The Case for Animal Rights, (Los Angeles: University of California Press, 1983).

39 For an analysis of environmental ethics that is predicated on natural rights, see Roderick Nash, The Rights of Nature, Leichhardt, NSW: Primavera Press, 1990. Nash argues that natural rights (beginning with Locke) have been ascribed to progressively wider classes of being and that this progressive ascription of natural rights can validly be continued. He says we are currently at the stage of ascribing rights to sentient animals and that ethics is likely to evolve to ascribe rights progressively to plants, life itself, rocks, ecosystems, the planet and the universe. None of the theories discussed in the present chapter recognizes a need for ascription of rights, although it should be noted that they do not preclude rights being ascribed to achieve political ends.

The assertion that an entity has what Taylor calls *inherent worth* entails two moral judgments: firstly, that the entity is a moral subject and therefore deserves consideration, and, secondly, that 'all moral agents have a *prima facie* duty to promote or preserve the entity's good as an end in itself and for the sake of the entity whose good it is'.[40] The concept of inherent worth applies in a human ethic to persons and in an environmental ethic to individual living entities in the environment. Its application, Taylor suggests, has four main implications in human ethics and four strictly analogous implications in environmental ethics. The implications for environmental ethics are that: if any wild creature has inherent worth then all have it equally; each individual should never be treated as a mere means to a human end; the promotion or protection of each one's good is taken as the ultimate end; and moral agents are required as a matter of principle to give consideration to the good of such a being.

This distinction between a human ethic of respect for persons and an environmental ethic of respect for nature seems, however, to be arbitrary and unnecessary. Humans, after all, are individual living things and, in common with all other living things, have inherent worth (as Taylor describes it). Humans are part of Nature and so having an attitude of respect for Nature would entail having an attitude of respect for humans. If all entities that have inherent worth have such worth equally, as Taylor maintains, then why should there not be just one ethic that covers the relationship between moral agents and all moral patients (or morally considerable entities)? And besides, not all humans are necessarily moral persons (for example, under Feinberg's criteria described below) and it is certainly possible that not all moral persons are humans—if, for example, we accept the premises of the Great Ape Project. But what is a 'person' such that it can or need be differentiated from other living things?

40 Taylor, Respect for Nature, p.75.

Not all cases of moral personhood are clear-cut. Feinberg, for example, argues all and only moral persons *actually possess* a set of defined characteristics: e.g. consciousness, a concept of self and some self-awareness, minimal rationality, ability to make plans and act on them, ability to communicate.[41] Feinberg restricts these characteristics to humans, but not even to all humans. Those who are brain dead or irreversibly comatose are not moral persons for Feinberg. But, as Goodpaster points out, Feinberg's restriction of moral considerability to certain classes of humans is probably inconsistent.[42] Goodpaster does, however, admit that he might be misinterpreting Feinberg by making Feinberg's notion of 'interests' equivalent to his own notion of 'moral considerability'. But it is clear, even from Feinberg's own arguments,[43] that there are, potentially at least, non-human persons and this blurs the distinction between an ethic for persons and an ethic for nature. But just how blurred does the distinction become? The potential admission of other species into the genus Homo, for example, poses interesting questions about the moral status of members of these species and highlights the arbitrariness of the division that Taylor makes between *persons* and *nature*. Are the apes to be covered by an ethic of respect for persons, or are they to be covered by an ethic of respect for nature? On this point the Great Ape Project is unequivocal: apes are moral persons. The Great Ape Project, which has the support of many distinguished scientists, philosophers and others from across the world, is an organization whose aim is to have all the great apes included in the category of *persons*. The basic

41 Joel Feinberg, 'Abortion'. Matters of Life and Death, Ed. Tom Regan. 2nd Ed. NY: Random House, 1986.

42 Kenneth E. Goodpaster, 'On Being Morally Considerable'.

43 Joel Feinberg, 'The Rights of Animals and Unborn Generations,' W. Blackstone ed., Philosophy and Environmental Crisis. Athens, Ga: University of Georgia, 1974. Cited by Goodpaster.

idea behind the project is to grant the nonhuman great apes the basic moral and legal protection that only human beings currently enjoy and its long-term goal is a United Nations Declaration of the Rights of the Great Apes. There are a number of internet sites related to the Great Ape Project.[44]

In purely biological terms it can be argued, as Rodd[45] does, that the great apes (gorillas, chimpanzees, orangutans) should be considered non-human members of the genus Homo on the basis of strong genetic similarity. The chimpanzee (*Pan troglodytes*), for example, has 98 per cent of its DNA in common with humans (*Homo sapiens*) and

> by analogy with similar groupings of genera of, say, bears, or elephants [one would have to say] that there is not one living species of 'man' which may be contrasted with 'animal' species, but four (*Homo,* two species of *Pan*, and *Gorilla*) or perhaps five if we consider that the orang's membership of the Hominidae is a sufficient qualification.[46]

Rodd goes on to suggest, on the basis of research into chimpanzee social group behaviour, that the great apes are conscious, have self-consciousness, can make plans, have language etc. The Gardners' work with the chimpanzee named Washoe indicates that chimpanzees can use, modify and pass on to their progeny human sign language. Whether this is also evidence that chimps have a language of their own is a matter of controversy. Lawrence Johnson discusses Washoe and notes that the chimp's ability to communicate with humans in a human language (American sign language, Ameslan) 'seems to involve an *additional* morally significant dimension...Washoe...as a language user...in some way at least [is] an associate member of the human tribe... she [Washoe] seems to be a peculiar kind of just-about-

44 www.greatapeproject.org

45 Rosemary Rodd, Biology, Ethics and Animals, Oxford: Clarendon, 1990. pp. 33-37.

46 Ibid., p. 35.

human being, with some associated moral status'.[47] Johnson says that to exclude beings such as Washoe from moral considerability is to operate on the basis of the Cartesian assumption that only humans are moral beings. One could speculate on the basis of the work with the great apes, using Feinberg's criteria that the apes have at least a *prima facie* claim to be moral persons. And if the apes can be considered to meet the minimum criteria for moral personhood, as supporters of the Great Ape Project contend, what of the cetaceans and other, terrestrial, mammals such as elephants? Having an ethic of respect for persons alongside an ethic of respect for 'nature' sets up a conflict that may be difficult to resolve. And, in any case, the above considerations suggest that such a distinction between a nature ethic and a human ethic should not be made at all.

Conflict and Resolution

Taylor maintains that there is a symmetry between a human ethics based on respect for persons and a life-centred ethics based on respect for nature. Despite the fact that it is possible to have a nature ethic based on respect for persons by extending the definition of moral personhood to sentient animals, ecosystems or even the Earth itself (as Gaia), Taylor maintains that the two systems of ethics are distinct, with each having respect as an *ultimate* end and that neither system can subsume or override the other, so that, for example, when a human's needs come into conflict with the needs of a living non-human the human's needs do not necessarily override the non-human's—and vice versa. In the case of human ethics the ultimate end is respect for persons whereas for environmental ethics the ultimate end is respect for nature. However, it is not at all clear why there

47 Johnson, A Morally Deep World, pp. 27-31.

should be two systems. Moral personhood should not be restricted to humans and 'nature' cannot be restricted to the 'wild' or even to the non-human world.

In (the inevitable) cases of conflict between Taylor's two systems the attitude of respect for nature cannot be subordinated to the attitude of respect for persons (to the degree that 'natural' entities are not persons). And neither should the attitude of respect for persons be subordinated because both are ultimate moral attitudes. So what are moral agents to do here? Taylor offers a hierarchy of duties to enable conflicts to be resolved in a manner that will bring the least pain to the respective parties and suggests that there are five principles for the fair resolution of conflicting claims and a hierarchy of four rules of duty by which different moral weights are assigned to different types of duty. There are difficulties with Taylor's principles, but it should be noted that for any ethic to be practically useful it needs to be able to resolve cases of conflicting interests. Taylor appeals to a comparison of the individual characteristics of morally-considerable entities, but this does not take account of the value an individual has to the system of which it is part. A more inclusive strategy would be to use supervening ecosystemic properties as the basis of differentiation between individuals. Accepting that there are supervenient ecosystem properties or characteristics and that these are morally relevant implies that there is no one right or correct way to treat any given individual as how an individual is treated will depend on how it is positioned in its environment. A short-hand way of describing this is to say that judgments are made on the basis of the niche value[48] of individuals or

48 The notion of 'niche' was introduced into the environmental sciences by Joseph Grinnell in 1913 to denote the role that a species plays in the local ecology. In his 1927 text Animal Ecology, Charles Elton wrote 'It is...convenient to have some term to describe the status of an animal in its community, to indicate what it is doing and not merely what it looks like, and the term used is 'niche'...the 'niche' of an animal means its place in the biotic environment, its relations to food and enemies' (Charles Elton, Animal Ecology, Oxford: Oxford University Press, 1927, pp. 63-4. Cited by Peter J. Bowler, The Fontana History of the Environmental Sciences, London: Fontana, 1992, p. 529).

on the niche value of kinds. A niche describes a function required of an ecosystem so that if the niche is filled the ecosystem continues to operate without major change to the relations which comprise the system. Niche value, therefore, is the value that an ecosystem (and hence the members of the ecosystem) has for the *function* performed by an organism. It does not matter what organism performs the function, as long as it is performed. If the function is not performed, the ecosystem changes and the members of the system are affected in some way. If a function with a high niche value ceases to be performed then the resulting changes to the ecosystem will have a deleterious effect on a high proportion of the members of the ecosystem. Some functions (and therefore the organisms that perform the function) have such a high niche value that if the niche is not filled, the ecosystem collapses. This ecological niche value will be joined by an ontological niche value and a semiotic niche value in later chapters.

But let's return to Taylor's five priority principles: a) the principle of self-defence, according to which it is morally permissible to do harm to others if it is the unavoidable consequence of protecting oneself from serious harm; b) the principle of proportionality, in which non-basic interests should never override basic ones (Another way to say this might be that a being's welfare interests should not be harmed in pursuit of mere preference interests of another being.); c) the principle of minimum wrong according to which, if harm to a being is unavoidable, then the harm inflicted should be the least possible; d) the principle of distributive justice, which consists in the claim that there should be fair distribution of welfare among interested parties; and e) the principle of restitutive justice according to which, if a wrong has been perpetrated, there is an onus on moral agents to make reparation for that wrong. Principles b, c, d and e are interrelated and cover cases where self-defence is not needed. In Taylor's words:

> The principles of proportionality and minimum wrong apply to cases in which there is a conflict between the *basic* interests of animals or plants and the *nonbasic* interests of humans. The principle of distributive justice...covers conflicts where the interests of all parties involved are *basic*...the principle of

restitutive justice applies only where, in the past, either the principle of minimum wrong or that of distributive justice has been used.[49]

Taylor's four rules of duty are a) the rule of non-maleficence, which is a negative duty i.e a duty *not* to inflict needless harm on a moral subject; b) the rule of noninterference, also a negative duty, in which moral agents ought not to interfere with the interests of moral subjects unless there are compelling reasons to do so; c) the rule of fidelity or non-deception under which moral agents ought not lie or engage in deceptive practices; and d) the rule of restitutive justice, under which a moral agent has a duty to restore the balance of justice when a moral subject has been wronged by the moral agent. Of the duties, the duty of nonmaleficence takes priority. Where conflicts cannot be avoided the priority principles are:

a) fidelity and restitutive justice override noninterference when a great good is brought about and no creature is permanently harmed...b) restitutive justice outweighs fidelity when a great good is brought about and no serious harm is done to a creature whose trust in us is broken.[50]

Trust here is the key element of non-deception. Taylor uses a number of examples, among them that of a fisherman who uses a lure to catch a fish. The fisherman sets out to deceive the fish, which as a moral subject, is entitled not to be deceived. The relationship of trust between the moral agent (the fisherman) and the moral subject (the fish) is broken by the deception, although one might ask just what relationship of trust exists here that could be broken. There is clearly no reciprocity of trust.

For any of Taylor's duties to be implemented, the individual moral agent must first recognize the obligation to take *responsibility* for moral subjects or patients, where taking responsibility is a personal virtue that individual moral agents can practise. An argument in

49 Taylor, Respect for Nature, p. 263.
50 Ibid., p.197.

favour of responsibility as the prime virtue for moral agents is outlined in Chapter 6. But let us look at some of the problems with the implementation of Taylor's scheme.

One of the main problems is that Taylor's theory requires certain important criteria to be specified independently of the ethic itself: for example, the self-defence principle. If I were to kill a police officer who was doing his or her duty in trying to stop me committing a crime, I could not legitimately plead self-defence. However, if I were to be killed by someone I was trying to mug, then that person could legitimately plead self-defence. Legitimate appeals to self-defence require that the defender be in the right. At the point when the principle of self-defence is introduced into Taylor's account there is no independent criterion for deciding which agents (human or non-human) are in the right and which agents are in the wrong. In fact it may not be possible for there ever to be a single criterion that is applicable to both the (putative) antagonist and the defender because there is no all-embracing notion of right and wrong that can be applied. If I were to shoot a lion that was about to attack me, or kill a snake that was about to strike, I could not legitimately claim self-defence in the same way as that described above because to do so would require that I be in the right (and about to be wronged) and the lion or the snake be in the wrong. Neither the lion's nor the snake's viewpoint is available to me and mine is unavailable to them. The viewpoints are incommensurable. I would still defend myself, but it would be categorically different from the situation where I was wronged by a moral agent. Taylor's principle of self-defence is not completely coherent. In the same way that the self-defence principle requires (but does not have) an independent criterion for deciding who or what is in the right, so, too, does the principle of non-maleficence require for its coherent application a notion of not doing harm that can be supplied independently of the principle. Moreover, the other principles that Taylor puts forward are open to the same form of attack.

A further problem for Taylor is that he applies his own principles inconsistently. In the latter part of his book, for example, and against his own stated principles, he licenses quite serious infringements of the vital interests of other living things in pursuit of non-vital and relatively trivial human interests. It would seem that—contrary even to his own claims—Taylor takes a position that is implicitly anthropocentric without supporting argument.

Moral Significance and Merit

When Taylor extends the range of morally considerable entities to cover all biological organisms, he makes all entities equally morally considerable. But this does not of itself provide any guidance as to how to decide which entity should have priority treatment in instances of competing valid claims, and his own principles for doing this have been shown to be lacking. So, what other ways might there be? Goodpaster's notion of moral significance seems *prima facie* to allow a ranking of competing claims that may be consistent with a ranking method based on the supervenient ecosystemic properties of individuals (or species). Moral significance, says Goodpaster, 'aims at governing *comparative* judgments of moral 'weight' in cases of conflict' and is a separate question to that of moral considerability. The question of whether a tree is morally considerable is separate to the question of whether dogs deserve more or less consideration than trees or whether humans deserve more consideration than dogs.

Relative moral significance may be determined, in part if not in full, by the function of the entity in ecological terms. This seems a good way to deal with relative moral significance because it takes into account how a change in one (living) thing will affect the interests, particularly the welfare interests, of the other living things with which it shares an ecosystem. Moral significance spelled out this way does not, however, provide a stable value but one that will vary with the status of the system as a whole. A system threatened by, for example, a rising water table and consequent salinification will value trees that can help keep the water table from rising. When the water table is lower, however, these same trees will be relatively less valuable to the system as a whole. It needs to be remembered, however, the value that a system has for a niche function is not a moral value. Moral values enter such systems with moral agents who can use niche values to make judgments that determine what actions should be taken.

This ecological function basis for making judgments appears to be consistent with Taylor's concept of merit. A pragmatic decision may be made to privilege one morally considerable entity over another because, in terms of the question being addressed, it has greater *merit*—its value to the ecosystem of which it is a member is greater than the value of the other. Merits are good-making properties and defects are bad-making properties. Merit cannot be used to differentiate the moral *considerability* of one entity from another. They may, though, have different moral *significance*. The degree of merit an entity has depends on the point of view from which merit is assessed. For example, from a human point of view, in Australia (where I am writing this), the fox has merit as a predator of rabbits because rabbits are an economic pest. The fox has a defect because it is also a predator of commercially-valuable stock and indigenous fauna. From a non-human viewpoint, such as that of a large herbivore (say a kangaroo) a fox has merit because, by keeping rabbit numbers down, the pressure on a large range of flora species (that are food for the kangaroo) is reduced. Rabbits are severely destructive of flora and this destruction has a deleterious effect not only on the flora but also on other herbivores—wild or domesticated. From the viewpoint of small indigenous marsupials, such as the endangered western quoll, the fox (as a predator of them) has a considerable defect.

Just as the function of an entity can be considered an ecologically supervenient property of the entity, merit also might be said to supervene on the individual's situatedness in an environment. Merit is not an essential part of any individual: it will vary according to which niche an organism fills in an environment and from whose perspective the 'niche' is viewed. One can choose between different beings on the basis of their respective merit, but one has to beware of applying to one category of existing things the criteria of merit that are properly applicable to another.

A member of a keystone species such as, say, the cassowary in northern Australian rainforests, would have greater moral significance than other species because if it disappears the ecosystem to which it belongs will collapse. (A keystone species is one that if removed from a system results in the virtual disappearance, directly or indirectly, of several other species, causing a chain reaction of species extinctions.)[51] The importance of the species—i.e. the sum of the values of its ecologically supervenient properties—determines for moral agents the relative significance of the individual members of the species. Significance here is related to the niche value of a species: the effectiveness with which a species, or individual members of species, fill an ecological niche. A cassowary in a northern Australian rainforest has a great deal of merit because the cassowary fills an essential niche in its ecosystem. Without the cassowary a great many other entities with a good of their own would suffer. Humans on this basis might well be considered to have little merit for most of the ecosystems of which they are a part i.e. if they are not removed the ecosystem will deteriorate (the welfare interests of system members will suffer) and may collapse.

Conclusions

Taylor adopts a form of biological individualism as one of the foundations of his environmental ethic—a position that can be defended—but his individualism needs to be supported by a clearer articulation of the nature of individual biological unity. Such a clearer

51 On the idea of keystone species, see particularly the work of E.O. Wilson.

articulation can be provided by Gill's discussion of unity in Aristotle. There are some problems with the Kantian ethic that Taylor superimposes on to this individualism—such as the subordination of feeling to rationality. There are problems also with the ability to resolve the inevitable conflicts that arise when two distinct ethics, a human ethic and a nature ethic, are allowed. These problems will be addressed further in Chapter 6 where the questions of responsibility with respect to an environmental ethic will be addressed and where an ethic which depends on a moral agent taking responsibility for moral subjects is developed. This ethic retains Taylor's idea that individual living things are teleological centres of activity and can be thought of as moral subjects, but avoids dividing ethics into an ethic for 'persons' and an ethic for 'nature'—and avoids excluding that part of 'nature' which is not 'wild'.

Freya Mathews takes a much wider view of what 'nature' is and of what is morally considerable, than does Taylor, so the next chapter examines her theory of the ecological 'self' and her use of what she claims to be the Spinozist concept of *conatus.*

5. Selves, Conatus and Aristotle

Introduction

In seeking to establish the grounds for an ethics we should ask what is the key element (or elements) of an entity that makes it a candidate for moral considerability. So far in this book there have been discussions of three concepts that have a claim to be the key element the possession of which bestows moral considerability: *conatus*, autopoiesis, and the biological teleology of Aristotle associated with his concept of *dūnamis*. The discussions have shown that there is much in common between these concepts. Each in its way refers to living individuals; each attempts to explain what it is that allows a thing to persist as itself over time despite changes; and each refers to a force *internal* to individuals—that is, in each case there is an immanent teleology or internal directedness operating. So, rather than there being three concepts, will one suffice?

From the previous two chapters it is clear that both an autopoietic ethic and an ethic based on the concept of individual living things being 'teleological centres of activity' can be expressed substantially in Aristotelian terms. And it will become clear that an ethic placing a centrally-important emphasis on *conatus* may also be expressed in Aristotelian terms. From there it is a short step to argue that an Aristotelian biological ontology is broad enough to be applicable to life-based ethics in general. It may be a short step, but it is important that it not be a hasty one. There is still a question of underlying metaphysics to be resolved, and in particular the question of whether a holistic metaphysics is the most appropriate to a full understanding of value in and responsibility for the natural world. A holistic metaphysic, in various forms, has underwritten a number of approaches to the natural environment —many of which have been grouped under the loose rubric of Deep Ecology, in which the concept of *conatus* has played a significant role. For example see Arne Naess's use of a Spinozist conception of *conatus* as a key concept in his environmental philosophy —a philosophy that was instrumental in creating so-called Deep Ecology.

If the concept of *conatus* can properly be said to be subsumable (in the context of an environmental ethic) by Aristotle's *dunamis,* there is a need to show that an ethic based on the concept of *conatus* may be expressed, without significant loss, in terms of an Aristotelian teleological biology and not be tied necessarily to a metaphysics such as holism or monism. To do this it will pay to examine Mathews' cosmological approach to environmental philosophy, at the heart of which is a version of *conatus* that she says is derived from a Spinozist view of the concept. It will also pay to compare selected areas of Mathews' theory with elements of Taylor's to show that both theories, which are examples of life-based ethics, have a significant debt to Aristotelian notions, particularly those associated with Aristotle's teleological biology.

We begin with Mathews' cosmological approach which, on the face of it, seems inconsistent with an Aristotelian position, then examine how she deals with the question of individuation before arguing that her use of *conatus*, which is the principle of individuation in her cosmology of substance monism, is closer to the Aristotelian archetype *dūnamis* than it is to Spinoza's concept of *conatus.* From this it will be clear that an Aristotelian teleological biology is a sufficient basis, or at least a sufficient ontological basis, for an environmental ethic.

A Cosmological Approach

In *The Ecological Self* Freya Mathews attempts to map out a cosmology for ecological ethics based on four main ideas: substance monism; an off-shoot of Einstein's relativity theory called geometrodynamics (GMD); a systems-theoretical notion of individuation; and a modified Spinozist conception of *conatus*. Her cosmology of substance monism, in which there is one universal substance, is derived from Spinoza's monism but also relies on the theory of geometrodynamics, in which 'space-time' is the single universal substance and all apparently material things are like 'knots' in it. Different material things are 'knots' of differing densities: water, for example, is a loose 'knot' relative to something like granite. For J.A. Wheeler, who first set out the theory of GMD, a geometrodynamical universe is 'a world whose properties are described by geometry, and a geometry whose curvature changes with time—a dynamical geometry.'[1] Combining a Spinozist approach with a version of modern physics provides Mathews with a 'way of understanding individuality in a context of interconnectedness'.

Substance monism is a direct challenge to the atomism which still functions as a strong ideological force in Western society despite the advent of quantum mechanics and relativity theories. Although contemporary physics has undergone a paradigm shift away from Newtonian mechanics, this paradigm shift has not yet been fully transferred into popular consciousness and mechanistic atomism still forms a major part of Westerners' worldview. Modern atomism—which Mathews wants to take issue with—was developed in the context of a mechanistic approach originating with Kepler's establishment of measurement and mathematics as the basis for investigating nature. In attempting to quantify nature Kepler broke with the Aristotelian system where Nature was seen in terms of forms, functions and essences. Other challenges to the ideology of atomism have also been made in the environmental literature, but usually by invoking some form of eastern metaphysics such as Taoist holism or

1 J. A. Wheeler, 'Curved Empty Space Time As the Building Material of the Physical World: An Assessment' in E. Nagel et al. eds. *Logic, Methodology and Philosophy of Science*, Stanford: Stanford University Press, 1962. Cited by Mathews, *The Ecological Self*, p. 61.

Buddhist interconnectedness.[2] Mathews begins her challenge from within the Western tradition by tracing the origins of mechanistic atomism through to the 'corpuscular atomism' of Newtonian physics. The Newtonian worldview promulgated a view of nature as something 'dead'—despite Newton's apparent dissatisfaction with mechanism and his view of life and will as 'active principles by which we move our bodies'.[3] Mathews notes, however, that an atomistic ontology need not be mechanistic, and that it is logically possible to 'postulate particles which embody a principle of dynamism, [particles] which are self-starters and self-movers'.[4] That is, for Mathews, it is possible for there to be ontologically distinct unities that have an internal *telos* or purposiveness —following Aristotle this could be called a *dūnamis*, or principle of dynamism.

The notion of a 'dead' nature such as that promoted by a Cartesian worldview had, says Mathews, a number of implications, the first of which was a move away from a notion of nature as something 'richly informed with *telos*, and with principles of spirit and agency'.[5] That is, a move away from a view of nature influenced by Aristotelian physics and metaphysics. For Mathews, the particular 'scientific' view promoted by a Cartesian worldview negates any 'sense of responsibility for, or identification with, Nature'—a sense of responsibility or identification which she argues is highly desirable on

2 On Taoism see, for example, Fritjof Capra (1983), The Tao of Physics. London: Fontana, 1969 or Graham Parkes, 'Human/Nature in Nietzsche and Taoism', J.Baird Callicott and Roger Ames eds., Nature in Asian Traditions of Thought, NY: SUNY Press, 1989. On Buddhism see, for example, pieces by Gary Snyder and Robert Aitken Roshi in Bill Devall and George Sessions ed. Deep Ecology: living as if nature mattered, Salt Lake City: Gibbs Smith, 1985.

3 Carolyn Merchant, The Death of Nature. Cited by Mathews, Ecological Self, p. 30. Newton's view of life and will as active principles of self-movement, i.e. as internal dynamic principles, arguably reflects elements of Aristotelianism.

4 Mathews, Ecological Self, p.17. Leibnizian monads might be seen this way, also.

5 Ibid., p. 31.

moral grounds.[6] Identification with nature is a central part of Taylor's thesis and Naess's and has normative implications in both these theories and Mathews'.

The second implication of a 'dead' nature relates to attitudes to self. Mathews claims that 'from a mechanistic perspective, Nature in itself is devoid of value... [which is] tantamount to saying that the fact/value dichotomy is built into the mechanistic worldview.'[7] This, however, overstates the case somewhat since a mechanistic view can allow for value by naturalizing it or emphasizing its supervenience on the material. She goes on to suggest that value enters the world with the emergence of living things because living things have an interest in what happens to themselves—although this presupposes a particular view of value. This concept of having an interest is an important one and is examined below, but the moment let us concentrate on issue surrounding her cosmological position. For Mathews, the 'scientific' worldview that she sees as deriving from Descartes and Newton embodies the distinction between fact and value and portrays the world as inert, insensate, devoid of *telos*, of value, purpose and meaning. She suggests that the modern world stands in need of 'cosmological rehabilitation'—that it needs a new cosmology to help it overcome the prejudice against living things inherent in the Newtonian/Cartesian worldview and to counteract its ideological force. The rehabilitatory cosmology Mathews puts forward is the monistic metaphysic described briefly above. This monistic metaphysic, she says, is 'sanctioned' by science: at the micro-level by quantum mechanics and at the macro-level by relativity theory, each of which 'speak of a universe which is one, whole and indivisible, a universe which is not a mere aggregate of discrete individuals, but is itself, as a whole, the 'unit' of reality'.[8] This is a view consistent with Capra's,[9] for example, a view in which quantum mechanics and

6 Ibid., p.33.
7 Ibid., p.37.
8 Ibid., p. 50.
9 Capra, The Tao of Physics.

relativity theory both present an organic, holistic or ecological view which is, in essence, a systems view. Mathews stops short of complete agreement with Capra—she does not, for example, include in her model two elements which Capra takes into his: the mind-dependence of physical reality and the primacy of system or process over substance.

The third consequence of a 'dead' nature lies in the implications it has for our attitudes to others, for the structure of social relations and hence of ethics itself. In the Cartesian view that underpins classical science, as Mathews sees it, there is no social whole from which the individual draws any element of identity. She adopts a systems approach and a criterion of individuation that allows her to claim that there can be individuals, which she calls 'selves', within a single universal substance. Individuals exist in a social whole as systems within the larger system. These 'selves' have an immanent motive principle by which they are 'realized' or brought into substantial being. This 'selfhood', she says, is a form of individuation consistent with the metaphysic of interconnectedness provided by geometrodynamics (GMD). And although Mathews does not say so it is also to a large extent consistent with the Aristotelian account presented thus far.

Systems in Mathews' metaphysic exhibit two characteristics originally identified with quantum mechanics: they are not localizable in the sense that Newtonian atoms are localizable and they are intrinsically dynamic. The characteristic of intrinsic dynamism is a fundamental principle of quantum mechanics, just as the discussion in Chapter 1 showed it to be fundamental to an Aristotelian biological ontology. Mathews says 'the irreducible lawlessness of individual phenomena in the quantum domain...is an expression of *intrinsic activity*...this 'lawlessness' manifested in random fluctuations in the energy levels of individual systems is really an absence of *extrinsic* laws: no external force or cause is required to bring about perturbations or disturbances in the states of the systems.'[10] This seems on the face of it to be consistent with an Aristotelian view of intrinsic activity—that is, of activity which is not generated from external forces—except for the claims about non-localizability. If something is localizable it is an ontologically independent entity. Organisms in an Aristotelian metaphysic are localizable, they are ontologically independent and are substances. Mathews' 'systems' are not ontologically independent. Non-localizability, according to Mathews, explains how the true identity of systems cannot be given purely in terms of properties manifested at a given point in space and time.[11] However, while it is true that the identity of any living thing cannot be limited to a point in space and time, it is not necessary to take on board the criterion of non-localizability, particularly if the identity of a living thing is viewed as the sum of its essential properties plus those properties that supervene on the thing because of its situatedness in an ecosystem.

Having sought support for her metaphysics at the micro-level of quantum mechanics, Mathews then argues that, at a cosmological level, geometrodynamics (GMD) provides the materials for a new worldview. Central to GMD is the idea that phenomena are merely disturbances in the structure of space-time, where space-time is the plenary substance. But, we might ask, as Mathews does, why space-time? Why should space-time be the framework of physical existence? In answering this she enlists the support of Spinoza, saying that what Spinoza says of substance and God is 'uncannily close' to what GMD ascribes to space-time and that Spinoza's argument explains why reality takes the form of a space-time continuum. However, in what follows it will be clear that that Mathews interprets Spinoza in such a way that her argument can be better expressed in Aristotelian terms — without substance monism.

10 Mathews, Ecological Self, p. 54.

11 Ibid., p. 54.

In the metaphysics of single substance monism that Mathews suggests be adopted, it is true by definition that the single substance cannot be individuated in substantival terms. However, if a systems view is adopted in conjunction with a metaphysics of substance monism (as Mathews does) then the single substance can be individuated in functional-systemic terms. On Mathews' own admission, however, there can be ontologically-distinct unities that embody a principle of dynamism, so why is a monistic metaphysic necessary? The simple answer is that it is not necessary and that there is an alternative that addresses the question of an individual's function, that recognizes that individuals have an immanent purposiveness and that takes into account the situatedness of individuals within a system. If individual living things are distinct and separate substances, as Aristotle took them to be[12], then the unity of any organism can be understood in substantival terms without an atomistic metaphysics. A living thing may also be understood, at least in part, in functional terms that take account of the entity's position within a system—without requiring a monistic metaphysics.

Each organism in an Aristotelian worldview has a place in an overarching system, or cosmos, which in part defines what that organism is, so it can only be understood as what it is in a complete sense when its situatedness in an environment is taken into account. (Situatedness may also be understood through biologist Jakob von Uexküll's notion of *Umwelt* or 'environment'. For von Uexküll (1864-1944), a living animal has an *Umwelt* that is not given to it, but is produced by it through physical, sensory, interaction with objects in the world.[13] The human *Umwelt* 'practically coincides with' Husserl's

12 Gill, Aristotle on Substance

13 Mathews, Ecological Self, p. 143.

Lebenswelt.[14] However, there is also an internal or collective Umwelt that explains how the various components of an organism interact and function together. Individuals are collections of internal *Umwelten* in which the components communicate with each other and maintain the individual as an individual. The individual so comprised —whether an insect such as a tick or a mammal or a bird—exists in a harmonious relationship that pre-exists any present individual. In Aristotelian terms, the essence (*to ti ēn einai*) of an organism is its prime function—its what-it-is-to-be. Its essence defines what it *is*. To understand this, it is helpful to distinguish two senses of function: the function that the thing has for itself and the function that it has for others in its biological community. This can also be treated as a distinction between an essential, self-reflexive function and, what might be called, an ecological niche function. Only the former is definitive of the thing itself, but both together comprise a complete picture of what the thing is. This view permits systems function to be included as part of the definition of what a thing is, but does not require that individuation be defined only in functional-systemic terms.

To understand Mathews' position on substance metaphysics, we need to understand what she means by substance. For her it has a double aspect: it must exist concretely, though not necessarily materially, and it has to be capable of existing in the concrete mode, as opposed to the abstract, independently of any other thing. In other words, a substance must be ontologically autonomous. However, the status of a substance as an ontologically autonomous entity is distinct from the idea of a logical individual (which a substance may also be). The idea of a logical individual is the idea of something capable of being treated as a logical subject of predication.

Mathews sets up two broad views of substance metaphysics: the atomistic and the plenic. In atomism there are infinite separate,

14 Chebanov, Sergey. (2001) 'Umwelt as life world of a living being'. *Semiotica*. Issue 134, pp. 169-184.

ontologically-distinct substances. In a plenic metaphysics there is only one distinct substance. In atomism the criterion of individuation is purely spatio-temporal and substantival. In plenism, because substance cannot be plural, the criterion of individuation cannot be substantival. So how can there be individuals apart from the single plenum? Mathews says that through intrinsic activity within the substantival continuum certain 'wave forms or constellations' create a distinction between ontological selves and others, although without them being separable from 'the substantival matrix...'[15] These 'wave forms' are cohesive collections of dynamically-related items that take on the form of ontologically distinct individuals. The distinct individuals so formed are self-realizing 'systems' within the plenum substance. For Mathews, the paradigm of the 'self-realizing system—or 'self'—is the organism.'[16] Under General Systems Theory (GST) the individuals with which we populate our world, epistemically, are systems, where, in Stafford Beer's terms, a system is any cohesive collection of items that are dynamically related.[17]

The systems view provides Mathews with a principle of individuation that allows her to have a monistic metaphysic yet still have individuals. The principle of individuation in Mathews' systems view has four parts, such that an open system which exhibits [1] self-regulation, [2] homeostasis, [3] equifinality and [4] goal-directedness is an individual.[18] But one can ask whether conceptions of substance metaphysics should be limited to the two broad views offered by Mathews. Aristotle, for example was neither an atomist nor a monist (although he did not entirely rule out the possibility of substance

15 Mathews, The Ecological Self

16 ibid., p.97

17 Ibid., citing Stafford Beer, Cybernetics and Management, NY: Wiley, 1959.

18 Equifinality is a concept from von Bertalanffy who recognized that 'vital phenomena', that is living things, may reach the same final state from different initial conditions and in different ways.

monism)[19], but took a position of substance pluralism in which individual organisms, including the simplest plants, were among the primary substances. The unity and autonomy of individual organisms guarantees that they are conceptually and ontologically independent.[20]

In Aristotle's scheme the paradigm case of an individual is an organism—which is also a primary substance. However, for organisms to be considered as primary substances, the so-called 'paradox of unity' has to be resolved—i.e. the question of how a composite of matter and form (which is what an organism is) can be considered a unity when the pre-existing matter from which it was generated remains within it. If pre-existing matter remains within the unity then it is not ontologically autonomous, i.e. it has to rely on something else for it to exist, and does not meet the fundamental criterion of substance. As discussed in the previous chapter, under the Aristotelian solution to this paradox organisms can be primary substances and yet still be composites. The cause of the unity of organisms, as primary substances, is an immanent dynamism or motive force—an active *dūnamis*. This active principle is vitally important for the self-conservation of individuals in the Aristotelian scheme. As Gill says,

> Given the vital role of the active principle in conservation [of unity], there is a decisive line to be drawn between those entities whose active principle is internal and all others. Inanimate bodies, with the exception of the elements, depend on external movers both to express their capacities and to avoid degeneration. Living organisms are truly primary substances because they are autonomous self-preserving systems. That they cannot finally resist deterioration is their tragic weakness.[21]

19 See e.g. Dougal Blyth, 'Wholes, Parts, and Sequences in Aristotle,' International Philosophical Quarterly, Vol. xxxiv, No. 4, Issue No. 136, December 1994, pp. 453-463.

20 Gill, Aristotle on Substance, p. 242.

21 Ibid., p. 241.

Like Aristotle's organisms, Mathews' 'selves' also are autonomous self-preserving systems that generally cannot resist deterioration. However, in the case of the cosmos, which Mathews says is a 'self', this is an open question. Mathews' 'selves' differentiate themselves within the one single substance through their own intrinsic activity, their own internal active principle. Aristotle's organisms differentiate themselves as individuals also by an internal active cause or principle. The paradigm of a 'self' for Mathews is an organism and the paradigm of an individual for Aristotle is an organism. The nature of the organism in the case of Aristotle is determined by the organizing form. In Mathews the 'self' is determined functionally and systemically i.e. by the function the self is to perform in the system. How then does Mathews' position differ in practice from an Aristotelian position? Mathews' answer is that her scheme recognizes an *ecological* ontology, whereas Aristotle's ontology is *biological.* She notes explicitly that the view she puts forward was 'anticipated by Aristotle,' but rejects his position because there is in Aristotle 'no anticipation of the concept of ecology, or of the indivisibility of systems'. She says Aristotle

> entertained a basically biological—though not ecological—ontology [in which] the world was composed of individuals whose unity was understood, after the fashion of the unity of organisms, in functional terms. Aristotle's metaphysic does not develop in a holistic or systemic direction, but stops short at an emphatic form of individualism—a pluralism of discrete biological individuals'.[22]

But since Aristotle did have a conception of the unity of the cosmos, and it can be argued that he also recognized a sort of proto-ecological view in the sense that he recognized that there was a good in the interrelation of individuals, the grounds for Mathews' rejection of an Aristotelian position need to be re-examined.

22 Mathews, Ecological Self, pp. 108-109.

As an example of Aristotle's proto-ecologism, consider the passage at *Metaphysics* XII (1075a12-20) where Aristotle says,

> We must also consider in which sense the nature of the universe contains the good or the supreme good; whether as something separate and independent, or as the orderly arrangement of its parts. Probably in both senses, as an army does; for the efficiency of an army consists partly in the order and partly in the general; but chiefly in the latter, because he does not depend upon the order, but the order depends upon him. All things, both fishes and birds and plants, are ordered together in some way, but not in the same way; and the system is not such that there is no relation between one thing and another; there is a definite connexion. Everything is ordered together to one end; but the arrangement is like that in a household.[23]

While there are elements of what might be seen as a proto-ecological viewpoint in Aristotle, the primary difference between Mathews' position and an Aristotelian position on this point is that Aristotle's notion of interrelationship concerns the interconnection that obtains between individuals in so far as they are all dependent on the same external source, i.e. a prime mover, whereas Mathews is committed to interrelation within a single substance. For there to be interrelation within a single substance there needs to be an account of individuation that allows an *ecological* approach and which is consistent with a metaphysic of substance monism. This account of individuation Mathews finds in the notion of *conatus*, discussed below.

Before moving on to Mathews' use of *conatus,* however, a further look at value in and of ecosystems is in order as although an ecological ontology may account for value in ecosystems, it is not the only way to do so and it frequently requires a significant metaphysical stretch that imputes singularity and possibly some notion of self-hood to a system. To avoid such a metaphysical stretch it is better to begin with a biological rather than an ecological approach, in which value is primarily vested in individual organisms rather than in things such as ecosystems. However, in beginning with a biological approach we still

23 Ibid., p. 110.

need to be able explain what value, if any, there is in ecosystems, so a brief diversion to look at ecosystem value is in order. Value arises in living individuals because they are self-realizing and determine their own ends. If there is value in ecosystems then it most likely a categorically different value to that inhering in individuals because ecosystems should not be considered ontological unities whose ends are immanently determined i.e. they do not have an immanent *telos* toward which they will normally proceed. It is possible to recognise value in ecosystems while retaining a biological approach by using the notions of essence and supervenience explained earlier. In an approach that uses these (essence and supervenience) the value that an individual has *in itself* and *for itself* does not change—although the value that an individual has to an ecosystem (i.e. its value to other organisms) will change with circumstances. The changing value might be called the organism's *niche value* and its value to itself its *intrinsic value.* Both are important. However, intrinsic value is a widely misused concept, having at least three different senses, as John O'Neill points out,[24] and is frequently the object of equivocation. He distinguishes between a) intrinsic value as a synonym for non-instrumental value; b) to refer to the value an object has in virtue of its 'intrinsic properties' and c) intrinsic value as a synonym for objective value.

Here, O'Neill's first sense is used. That is, an object has intrinsic value if it is an end in itself. Such intrinsic value is a first, and necessary, condition of moral considerability. Niche or ecosystemic value, on the other hand, may be used to rank the total value of any one entity against others and so form the basis of pragmatic decision-making. One consequence of this is that in the event of some sort of environmental *triage* having to be implemented, entities with a higher niche value would have a stronger claim to be protected or preserved than those with a lower niche value: a keystone species—one in

24 John O'Neill, The Varieties of Intrinsic Value, in A. Brennan ed., The Ethics of the Environment, Aldershot: Dartmouth, 1995, pp. 55-73.

whose absence an ecosystem will suffer significant damage—would have a higher niche value than other species in the ecosystem because of the potential impact on other intrinsically-valuable entities and so have the strongest claim to be protected.

Conatus

For Mathews the immanent activity by which a self maintains itself is effectively equivalent to the Spinozist concept of *conatus,* a concept she employs in part to argue that the cosmos as a whole is a self-realizing system. For her,

> the conatus is a principle, or form, which, while it may have causal origins, once established becomes a force or drive which enables individuals to resist causal affronts to their integrity. The notion of the conatus...is effectively equivalent to that of self-realizability, except in so far as the former lacks systems-theoretic connotations.[25]

But although Mathews says that her *conatus* was inspired by Spinoza's, the concept of *conatus* that she uses is not exactly that which is central to Spinoza's *Ethics*. In her conceptualization, *conatus* inheres in self-realizing (i.e. living) entities and on this basis it might be argued that her concept picks up on Spinoza's 'psychic' sense of *conatus* rather than *conatus* in its complete sense. In this it bears a closer relation to Cicero's *conatus* than to Spinoza's.

Given that Aristotle's immanent dynamism (*dūnamis*) was historically influential in the development of the concept of *conatus* employed by Spinoza and that a Spinoza-inspired *conatus* is the principle of self-realization in Mathews' ontology, it is clear that there may be a strong correlation between Mathews' theory and an

25 Mathews, The Ecological Self, p. 110.

Aristotelian (immanent) teleology. So, can Mathews' position on 'selves' (and their moral considerability) be expressed without significant loss in Aristotelian terms? In short, yes.

There is at least one significant way in which Mathews' theory parallels an Aristotelian (biological, internal) teleology, and this relates to the modifications that she makes to the Spinozist concept of *conatus*. In Mathews' version, the focus of Spinoza's conception of *conatus* is narrowed so that it inheres only in self-realizing (i.e. generally speaking, living) entities, what she calls 'selves'—unities individuated within a single substance by their properties as systems.[26] Through this modification Mathews' concept comes closely in line with the notion of an active *dūnamis* and in particular with that active *dūnamis* through which living entities act on themselves *qua* themselves. In the model of *dūnamis* that Gill sets out, there are in effect two sorts of immanent motive principle, of which Mathews focuses on the second: (i) that which carries with it the sense of inertia, the tendency of things to remain as they are unless acted upon by some force, and (ii) that which carries with it a sense of a psychic (although not necessarily conscious or self-conscious) state, a sense in which a cognizing entity determines through its own internal processes actively to pursue the maintenance of its own identity. Mathews acknowledges that Spinoza identified *conatus* with the principle of inertia applicable to all things, animate and inanimate, but claims that when Spinoza referred to individuals he was generally referring to organisms. That Spinoza was generally referring to organisms is supported by Curley's observation that although Spinoza's use of the concept of *conatus* subsumes the concept of inertia, Spinoza chose deliberately to use a term (*conatus*) which suggests the presence of a psychic state.[27]

26 In the absence of the concept of conatus, Mathews' position looks remarkably like von Uexküll's, in which there is both an internal and external Umwelt, and where the former is a function of internal communication between parts.

27 Edwin Curley. ed. and transl. The Collected Works of Spinoza, Vol. 1, p. 657.

As shown earlier, Spinoza's notion of *conatus*, or immanent motive principle, can be traced back to Cicero who recognized that there was an impulse to self-preservation inhering in every natural organism.[28] For Cicero, this immanent impulse to self-preservation was part of what he termed Peripatetic philosophy. Cicero's use of the term 'Peripatetic' includes Aristotelian philosophy although in his characterization of what was Peripatetic Cicero did not always clearly differentiate what was Stoic in origin from what was Aristotelian and in this case the reference is probably to Theophrastus, Aristotle's student. The key point to note is that the concept of *conatus* originally included the recognition of an immanent motive principle or impulse in *all living things*. It was Spinoza who supplemented the concept of *conatus* so that under it was effectively subsumed the concept of inertia. In stripping from *conatus* the concept of inertia which Spinoza added to it, Mathews recreates the concept of *conatus* in a form very close to its Aristotelian archetype, *dūnamis*.

In Chapter 2 a distinction was made between an active and a passive *dūnamis*—and how this distinction between two sorts of immanent motive principle can be used to explain why Spinoza could recognize that living things had *conatus* while also recognizing that something like a rock had *conatus* too. When Mathews says that the concept of *conatus* is equivalent to her concept of self-realization, especially as it applies to the paradigm case (i.e. organisms), she appears to be using *conatus* in a sense that is equivalent to the internal *active dūnamis* inhering in living things. Mathews' use of *conatus* is a more specific application than the broad application in Spinoza.

It is worth making a point at this stage that the theory of social ecology, most strongly associated with the work of Murray Bookchin, also puts an immanent motive principle at the heart of its ontology and sees itself as part of a 'long teleological tradition extending from the ancient Greeks'.[29] Social ecology is a form of dialectical naturalism

28 Cicero, De Finibus, IV,7, S16.

29 John Clark, 'Introduction' to Part 4, 'Social Ecology', Zimmerman et al. eds. Environmental Philosophy, p. 345.

that 'proposes a principle of ecological wholeness that sees the entire course of planetary evolution as a process aimed at increasing diversity and the emergence of value.'[30] There is, according to John Clark, much in social ecology that it has in common with an Aristotelian approach and, although Bookchin tends to avoid using the term teleology, the theory affirms that 'the entire process of development of life and mind has 'directiveness,' and is a movement toward the greater unfolding of value'.[31] Social ecology is a form of *dialectical* naturalism because it sees all of reality as being in a continual process of self-development and self-transformation. In common with Mathews' approach, social ecology also interprets phenomena as inseparable parts of larger wholes, although it does not claim to have a cosmology of substance monism. A key element in social ecology's dialectical naturalism is that it holds human society to be part of nature and that human 'interaction with one another is interaction with nature and that our interaction with nature is continuous with our interaction with one another'.[32] This can in some ways be seen as consistent with the Aristotelian view of humans as *politikon zoon*, or political (social) animals. That is, humans are part of Nature and the way they interrelate is an extension of their nature. If humans were not political animals, they would not be humans—it is an essential part of their being. In this respect there appears to be much, *prima facie,* that social ecology has in common with environmental philosophies based on living organisms' possession of an internal teleology, but detailed discussion of this is outside the scope of this book. This very brief account of an aspect of social ecology does, however, provide further evidence of the central importance of conative and teleological concepts to environmental ethics.

In Mathews' version of a conative environmental ethics, an entity with an internally-determined purpose 'matters' to itself—it can be

30 Ibid.
31 Ibid., p. 345. Clark calls this directiveness a nisus.
32 Ibid., p.346.

harmed or benefitted, it can be said to have self-interest. In Taylor's terms, a teleological centre of value. This is a significant point of commonality between the two theories, but they differ on the important question of interests.

Interest

For Mathews, the defining difference between an organism and other individuals is that an organism is a ''machine' for self-realization' that, through its activity, defines its own interest or value, and so brings interest and value into the world. She writes: 'for an organism to exist is [for it] to possess self-interest...[and] the existence of the organism coincides with its purpose, for its purpose is to exist'.[33] That is, the purpose of a self-realizing 'machine' is self-referential and, because the interest and the purpose of an organism coincide, 'interest enters the world, ontologically speaking, in the shape of the living system'.[34] (For Mathews the term 'living system' can denote ecosystems and biological communities as well as individual organisms). With the appearance of systems-for-self-realization the ontologically new quality of being-for-itself enters the world, a quality that has the special quality of purposiveness. A categorically new feature of a being-for-itself is that it is what Mathews terms 'self-envaluing' in that a being with an interest in its own perpetuation is a being that values its own existence. This valuing, moreover, need not be a function of consciousness, or conscious awareness but, claims Mathews, is rather constituted by the being

> actively seeking to maintain itself in existence...self-realizing beings, in other words, embody value just as they embody *telos*. To be actively self-maintaining

33 Mathews, Ecological Self, p.101.
34 Ibid., p. 101.

> is...to be valuable to oneself. Such activity is the primitive form of value, the preconscious form in which value first enters the world.[35]

For an organism to seek to maintain itself is 'constitutive of its valuing itself'. That is, because being-for-itself is self-valuing, so value enters the world. It should be noted, though, that this value is not necessarily moral value and Mathews' explanation of the emergence of value *per se* offers no suggestions as to how moral value comes to be. However, as shown in the next chapter, moral agents have the capacity to recognize the unique value of living individuals and this places them (the moral agents) under an obligation to act to preserve or foster that unique value. For Mathews, the activity of self-preservation is constitutive of *agency*, although not necessarily moral agency[36] and the notion of being-for-itself is a 'package which includes the notions of *telos,* self-interest and agency'.[37] It is not necessary that the being-for-itself be conscious. It is only necessary that the being-for-itself (i.e. the moral subject) value itself by the act of self-maintenance. It is worth mentioning here, because of similarities to Jonas' position outlined in the next chapter, that this establishes a strong link between interest and value. Something with self-interest is actively self-maintaining and so is valuable to itself: i.e., with interest comes value. This value *for* oneself is the crucial component in Mathews' concept of *intrinsic value*, since, on her definition, a being that seeks to maintain itself is intrinsically valuable. The value that a being assigns to itself (consciously or not) is its intrinsic value. On this account, however, intrinsic value is not objective value since there is still a necessary link between value and a valuer.

The question of *interest* appears to be treated differently by each of Taylor and Mathews, and each, for different reasons, needs to

35 Ibid., p. 104.

36 For Taylor, such activity is a demonstration that the individual can be considered a moral subject, but the presence of such activity is not sufficient for the individual to be a moral agent. A moral subject is any entity that has interests.

37 Mathews, Ecological Self, p.104.

supplement their definitions of the concept. Mathews equates interest with value, and the having of an interest as a fundamental property of self-realizing systems. Taylor, on the other hand, holds that it is possible for a being to have a good of its own, but not have interests. Mathews appears to limit her concern to welfare interests (as opposed to preference interests), whereas Taylor uses two notions of interest. One of Taylor's uses of interest is precisely the same interest as Mathews'—effectively the self-interest of any entity capable of being benefited or harmed. Taylor's second type of interest is a property only of those entities that are capable of consciously aiming at ends or that can take means to achieve such ends. Only beings that can experience satisfaction, dissatisfaction, fulfilment or frustration, and that care about what happens to them, can have this second form of interest. This precludes from having this interest all living things that do not have consciousness or, if they do, lack the ability to choose from among alternatives confronting them. All forms of plant life and the simpler forms of animal life are thereby excluded from having such interests. For Taylor, 'something can be in a being's interest [i.e. self-interest] and so benefit it, but the being itself might have no interest in it...it might not even be the kind of entity that can have [this second type of] interests at all.'[38] It is possible, Taylor says, for a being to have an interest in something that does not promote its own (long-term) well-being. For example, a person may have an interest in smoking tobacco, even though smoking tobacco is undoubtedly not in the person's (long-term) interests.

One way to summarize Taylor's position would be to say that he recognizes that all living things have welfare interests, but preference interests are limited to those beings with both consciousness and the ability to make choices. This, however, runs the risk of skewing the argument as the potential for having preference interests may be in some significant way a determinant of moral considerability. In limiting preference interests to beings with (self-) consciousness

38 Taylor, Respect for Nature, p. 63.

Taylor implicitly denies an aspect of moral considerability to plants and the lower animals (although just how low is not clear). This could be seen as a contradictory position for Taylor, arguing as he does that all organisms are teleological centres of activity and are therefore morally considerable. Both Mathews and Taylor, then, have a deficient account of interests—the one for not recognizing preference interests as significant and the other for limiting arbitrarily the beings that might have preference interests.

Mathews' and Taylor's differing ontologies also lead them to apparently different positions on the question of what constitutes an individual and so of what constitutes a moral subject. It is worth comparing Mathews and Taylor more closely on this point—not just because they are prominent examples of, respectively, an ecocentric and a biocentric ethic, but also because, on closer examination, the apparent difference between them narrows considerably and by examining their positions we can come to a clearer understanding of what an individual is.

Individuals

There are strong similarities between Mathews' ecological ethics and Taylor's biological approach—for example, Mathews' self-realizing individuals are self-valuing entities with an internally-determined *telos* that might readily be seen in Taylor's terms as teleological centres of value; and for Mathews, as for Taylor, the paradigm case of an individual is a biological organism. But there are also significant differences. For example, Taylor's individuals are biological, material organisms whereas, although she also holds that organisms are self-realizing individuals, Mathews intends the category of self-realizing individuals to include other entities such as ecosystems. At a *practical* level, however, there would seem to be very little difference between Mathews and Taylor on what constitutes a morally-considerable individual since, even if an ecosystem is seen as a morally considerable individual, any change in it must be coeval with change in a constituent (and therefore ontologically prior) morally considerable entity. Also, as noted in the previous chapter, the concept of ecological supervenience in combination with the intrinsic value of organisms can explain value in ecosystems without ecosystems having to be distinct self-realizing entities. In short, there need not be an *ecological* individual.

In arguing the last point, however, clarity is needed on what is meant by 'ecological'. Taylor, for example, is firm that the science of ecology is not a basis for an environmental ethic. According to him, to claim that it is a suitable basis is to confuse fact and value—to confuse 'is' and 'ought', although whether fact and value constitute a dichotomy at all is contestable, as shown in the next chapter where the very existence of an organism as a good-in-itself introduces value into the world and so dissolves the distinction between is and ought, between fact and value. Taylor argues that to use the science of ecology as a basis for moral decisions is to confuse *is* with *ought* because although ecology can be used to explain how things are—that is to state the actual relations which hold between elements in an environment—it can imply nothing as to how moral agents ought to act or how human culture should fit into the order of nature. Taylor writes that there is nothing in the science of ecology that can tell us whether it is wrong to exploit the natural world.[39]

Mathews, however, does not use ecology or eco-centrism in the narrow sense of the science to which Taylor refers. She uses the term 'ecology' as a metaphor for interconnection in much the same way that Gregory Bateson uses it.[40] In fact, Bateson's 'mind' and Mathews' 'cosmic self' are in many ways identical. Ecology for Mathews thus refers to the idea of interconnectedness itself and

39 Ibid., p. 51. He notes that Aldo Leopold's A Sand County Almanac and Paul Shepard and Daniel McKinley's The Subversive Science make this mistake.

40 Gregory Bateson, Steps to an Ecology of Mind; Gregory Bateson, Mind and Nature: a Necessary Unity, NY: Bantam, 1988.

implies a particular metaphysic. For Mathews, however, the problem of moving from 'is' to 'ought' still remains. As explored in the next chapter, one way to avoid this problem is for each living thing to address an 'ought' to moral agents. That is, moral agents are placed under a duty of care and responsibility because a living thing is not a 'mere *is*', not a mere existent, but is an always already valued existent.

In maintaining that that there is a primitive form of value—that is the value or interest-in-self-realization that a self has for itself—Mathews aims to develop a theory which 'prescribes a particular normative stance to Nature'—a stance that is predominantly *ecocentric.* It is *predominantly* rather than purely ecocentric, because, although she begins her final chapter by saying that the normative stance to nature is ecocentric, she closes by saying that the normative thrust of her thesis is that we should adopt a bio-centric *or* eco-centric ethic.[41] Although Mathews claims that the distinction is not significant, it suggests, at least, that she believes that the practical results will be much the same regardless of which position is taken.

Taylor and Mathews take similar positions on the teleological status of living things. For Taylor, only a living thing can be a teleological centre of life, that is, have an internal purposiveness, or *telos*, that does not depend on an external agent. The internal purposiveness takes the form of each organism pursuing its own good. To quote Taylor:

> To say it [the organism] is a teleological centre of life is to say that its internal functioning as well as its external activities are goal-oriented, having the constant tendency to maintain the organism's existence through time.[42]

This need not entail consciousness or intention, so, for example, trees and other plants can be seen to have a good of their own around which their behaviour is organized. This position on teleology is one of the

41 Mathews, Ecological Self, p. 147.
42 Taylor, Respect for Nature, pp. 121-122.

four core commitments in Taylor's biocentric outlook. What he calls 'holistic' theories (such as Aldo Leopold's *Land Ethic*) give no place to the good of individual organisms other than as arising out of or related to their role in promoting the good of the ecosystem in which they live. Mathews' 'holism' is quite different, however, in that she does acknowledge a place for the good of individual organisms (as the paradigm of 'selves', her basic morally-considerable entities).

However, as argued earlier, there is no value in the ecosystem itself over and above the sum of values of the constituents: an internal *telos* inheres only in individual organisms and groups do not have a good of their own. A group may have use-value for any or all of its constituents, but in themselves the only good that groups (such as social groups or ecosystems) have is the sum of the goods inhering in the individuals that make up the group. Individuals have an intrinsic value and a 'niche value' that will vary according to the niches they fill and the efficacy with which they fill them. The niche value is a value *for* other constituents of the system.

Both Taylor and Mathews hold that teleology inheres only in individuals, it is just that they differ on what constitutes an individual: with Mathews holding that an ecosystem can be considered a 'self' or an individual with interests.

The ethic developed in the next chapter relies on the concept that a *telos* inheres in biological individuals and that this inherence is morally significant. But in this ethic, unlike Taylor's, having an 'attitude of respect' does not provide a moral agent with sufficient impetus to act in the interests of those entities which are to be 'respected' and ecosystems or ecological communities do not qualify as morally-considerable entities in their own right. The position in the next chapter derives from the Aristotelian biological ontology outlined above and relies both on moral agents recognizing that they are responsible for moral subjects and on those moral agents practising responsibility as a virtue.

Self-realization is, for Mathews, the major difference between mechanical systems and organisms. This can be seen in Aristotelian terms as the distinction between *technē* and *phūsis*. For Aristotle, only living nature (*phūsis*) is self-realizing. *Technē* requires an external moving force whereas *phūsis* does not. For Mathews, an organism is a 'system devoted to its own realization...the organism is a 'machine' for self-realization'.[43] The fundamental difference between this 'machine for self-realization' and other individuals, whether they are systems or substances, is that organisms define an interest—the interest or value they define being their own. This interest constitutes an 'immanent *telos*', a term that is centrally relevant to the concept of *conatus* that Mathews borrows (and amends) from Spinoza, and to Aristotle's biological teleology.

To summarize, Mathews' concept of self-realization (which relies on entities having *conatus*) is substantially the same as the concept of autopoiesis and both can be expressed in terms of the Aristotelian concept of *dūnamis*. And, given that Taylor's individualism can also be explained in terms of *dūnamis*, it would appear that there is a good case to treat an Aristotelian biological ontology as broad enough to be applicable to life-based ethics in general.

Mathews' and Taylor's theories are important representatives of, respectively, ecocentric and biocentric ethical positions. If there is a view that can embrace both of these positions without losing the strengths of each then it should provide a strong ground on which to base an environmental ethic. An ethic based on an Aristotelian biological teleology can indeed do this.

43 Mathews, Ecological Self, p. 97.

PART III:

VALUE AND RESPONSIBILITY

6. Value, Complexity and Obligation

Introduction

The ethic of responsibility developed in this and the next chapter relies on the idea that living organisms have an immanent *telos* and a good of their own. To this will be coupled the notion that the emergence of this immanent *telos* (which occurs with the emergence of living things) introduces value into the world and that the presence of this value imposes an obligation of responsibility on moral agents—a responsibility to protect, care for, nurture and help realize, the good-of-their-own that all living things have as a defining property. Accepting and exercising such responsibility is a virtue.

There are five aspects of an ethic based on an Aristotelian biological teleology that need in particular to be expanded and clarified. The first is the idea that, with the emergence of entities with a good-of-their-own, interest and value enters the world. The second is a claim that the relative complexity of entities is important, with more complex organisms having a stronger *prima facie* claim on moral agents than do less complex organisms. The third aspect is the notion that the obligation under which moral agents are placed is one of *responsibility*, where responsibility is an obligation that requires moral agents to *act* to protect the good of moral subjects and where all living things are moral subjects. The fourth is the idea that recognition of responsibility is not necessarily based only in a narrow form of reason, but is also rooted in feeling. And the fifth is that when this responsibility is recognised and the agent chooses to act (and so acts) the agent takes on the virtue of responsibility. The first three are discussed below, aspects four and five are discussed in the next chapter.

Individual living things bring value into the world because unlike non-living things each has internal directiveness and a good-of-its-own and is at the very least self-valuing—i.e. each has its own end within it, an end that it has an interest in achieving. But what is 'value' that it can enter the world? To acknowledge a feature of something as having value is to take that feature into account when making decisions. But this is an 'outside-in' view that does not address the issue of the value something has in and of itself. Value enters the world, and the existence of value becomes a fact of the world, with the emergence of internally-directed entities that value themselves. This is an intrinsic value: a categorically different sort of value from that which is sometimes taken into account when making decisions. The value that is taken into account in making decisions is commonly a *use* value which need not be recognized in the same way by all possible valuers. Intrinsic value may be taken into account when decisions are made, but the sorts of decisions involved will be those that affect or potentially affect the welfare interests of the entity that has this (intrinsic) value. This (intrinsic) value enters the world because each being with internal directiveness and a good of its own has an interest in maintaining that good and in pursuing its immanent goals, this interest in pursuing an immanent goal being a value to any being that can pursue such goals.

To summarise: with the evolutionary appearance of organisms—living things—the ontologically new quality of being-for-itself enters the world. Being-for-itself has the special quality of purposiveness (*telos*) and with the emergence of these new qualities, 'interest'—or self-value—enters the world. Each organism has an interest in maintaining itself as a living organism—an interest in realizing its immanent potential—and because of this it values itself. And because the being-for-itself values itself, value enters the world.

Why, though, should anything other than the possessor of the intrinsic value take that value into account when making decisions?

How do we move from value as a fact of the world to a prescription for action based on value? How does one move from 'is' to 'ought'?[1] One way is suggested by Jonas[2] who says the prime example intended to show that 'is' implies 'ought' is that of the newborn, which he takes to be the object of responsibility in what he calls the paradigm case of 'natural' responsibility. The newborn is not a *mere* 'is', a mere existent, but is a valued 'is' as soon as it enters the world. It has value-for-itself to be sure, just as any living thing has, but it also has a value for the parent (or other moral agent) who is obligated and has an 'ought-to-care' relationship with the child-as-object simply because the child 'is'. What makes the newborn 'the perfect paradigm, literally the prototype, of an object of responsibility...lies in the unique relation between possession and non-possession of being, displayed by beginning life, which demands from its cause to continue what has begun.'[3] The 'cause' here is presumably the newborn's parents, however it can refer also to other members of the species to which the newborn belongs. In the absence of the newborn's parents, other members of the species have a responsibility to the newborn. In the absence of either of these, any moral agent has such a responsibility because the child-as-object has a good that carries with it an immanent *demand* continually to be realized. In Jonas' words, the newborn's 'mere breathing uncontradictably addresses an ought to the world around, namely, to take care of him. Look and you know...the plain being of a *de facto* existent immanently and evidently contains an ought for others'.[4] Jonas uses 'uncontradictably' rather than 'irresistibly' because the force of any 'ought' can be resisted.

1 Rolston makes the observation that reasoning in environmental ethics is often a 'transition from is to good and thence to ought'. Holmes Rolston III, Environmental Ethics: Duties to and Values in the Natural World, Philadelphia, Temple University Press, 1988.
2 Hans Jonas, The Imperative of Responsibility, Chicago: University of Chicago Press, 1984.
3 Ibid., p. 131.
4 Ibid.

Jonas and Emmanuel Lévinas, who we will look at shortly, were both students of Heidegger and were both in various ways involved in taking issue with the nihilism at the heart of Heidegger's attack on metaphysics. They did so, however, while sharing Heidegger's contention that the 20th Century was a time of crisis for the West because modern reason under the sway of science was self-destructive and that it was the job of philosophy to show the limits of science by returning to ancient philosophy.[5] Both Jonas and Lévinas also adopt positions based on a notion of 'responsibility' in which the vulnerability of an other in some sense calls us into responsibility.

One of the functions of an ethic predicated on responsibility—such as the one outlined below and in the next chapter—is to educate moral agents so that they can learn to recognize those things toward which they have a responsibility. Although—despite their common heritage both as Heidegger's pupils and Judaic scholars—Jonas and Lévinas have markedly different metaphysical bases to their views,[6] Jonas's Kant-inspired 'imperative' of responsibility might, however, also be couched in terms consistent with Lévinas' view, such as when Lévinas says: '...There is a commandment in the appearance of the face, as if a master spoke to me...'[7]

An ethic of responsibility for nature makes the possession of certain forms of immanent purposiveness relevant, perhaps centrally relevant, in determining the moral considerability of an entity. Such an ethic also views humans as part of nature and not in any way apart from nature since in this ethic *immanent* purposiveness is a characteristic of all living things, including humans. That humans also

5 Lawrence Vogel, Overcoming Heidegger's Nihilism, in Samuel Fleischaker, ed., Heidegger's Jewish Followers, Pittsburgh: Duquesne University Press, 2008, p. 131.

6 Lawrence Vogel, Jewish Philosophies after Heidegger, in Winston Davis, ed., Taking Responsibility: Comparative Perspectives, Charlottesville: University Press of Virginia, 2001.

7 Emmanual Lévinas, Ethics and Infinity: Conversations with Philippe Nemo, Trans. Richard Cohen, Pittsburgh: Duquesne University Press, 1985, p.41.

have a *conscious* purposiveness may separate them off as moral *agents* but, *prima facie*, gives them no special privileges as moral *subjects*.

In summary, a living thing is a being-for-itself with an immanent good that places anyone who recognizes that good under an obligation to care for it. There may be a case for differential treatment of self-realizing entities, for giving one priority over another at a given time, but that is a different matter, on which discussion of the issue of complexity can shed some light.

Complexity is Relevant

It is intuitively satisfying that we should take a position in which the more complex an organism the greater the moral regard that is owed to it. But this may merely betray a very human bias that reflects a conception of our own complexity and our status as moral subjects. That is, we are complex entities and we are indisputably moral subjects, so we have a tendency to see that moral status is linked to complexity. But is there a link between intrinsic value, complexity and moral considerability? Why should these matter to moral status? Nicholas Agar[8] recognizes intrinsic value as relevant but notes that it is hard to attribute such value to non-sentient life forms and hard to demonstrate that not all sentient species should be valued equally. Under his account moral good is identified with an organism's 'biopreferences'—or environment-related goals (to do such things as feed)—and individual living things, and not species or ecosystems or communities, are the entities to which we owe moral concern. His account allows all living beings to be morally relevant, but does not

8 Nicholas Agar, Life's Intrinsic Value: Science, Ethics and Nature, New York: Columbia University Press, 2001.

ascribe equal moral value to all living beings. For him, how much value is assigned to a living thing depends in part upon the range and complexity of the goals of which the organism is capable. In presenting his account he does address the issue of complexity, but falls short on the details as to why complexity matters.[9] So, is a position that greater complexity implies greater moral worth defensible? This clearly depends on how we understand complexity and whether it is related to value. But before continuing with the relationship between complexity and value we should note that there are two areas of complexity relevant here. The first is the complexity of individual organisms and the second is complexity of the interrelations between an organism and its environment.

Each living thing has a value to its own environment and this should be considered when making decisions that might affect living things. That is, in addition to any intrinsic value an organism has it also has a value that varies according to the degree to which that organism interconnects with other organisms in the ecosystem. To the extent that the ability, capacity or potential to interconnect is 'built-in' to the organism, this capacity, ability or potential should be seen as contributing to value. The capacity to interconnect will be discussed below as part of ideas associated with Jakob von Uexküll's concept of *Umwelt.* But before moving to von Uexküll's ideas, it is useful to distinguish between the *potential* for interconnection, which (following Aristotle) is an unvarying and essential 'built-in' characteristic of an entity, and the *actual* interconnections an entity makes at any one time, which vary with circumstances. If the potential for interaction is a factor in the value of an organism this would lend weight to the claim that complexity is relevant in that a more complex organism, with its greater potential to interact, has a greater potential to affect the interests of other organisms. Complexity potentially allows greater interaction, but complexity alone does not guarantee

9 Jan Hancock, Review of Nicholas Agar, Life's Intrinsic Value: Science, Ethics and Nature, Australasian Journal of Philosophy, Vol 80, No. 4, December 2002, p. 528.

greater interaction—interaction depends also on the external (objective) world.

We will come shortly to discuss what is meant by interaction and how it may be understood in terms of the interpretation of signs. But for the moment, an organism's *actual* interconnectivity may usefully be seen not as part of any intrinsic value but as an ecological niche value. This refers in the first instance to particular, identifiable, individuals and secondarily to species, as different species may fulfil substantially the same niche functions in ecosystems, e.g. as scavengers, as predator or prey or as seed dispersal agents. This ecological niche value can be employed to differentiate between organisms and between classes of organisms. To use an example introduced earlier, a member of a rare 'keystone' species, such as the Southern cassowary (*Casuarius casuarius johnsonii*) in the northern Australian rainforest, is more valuable to the ecosystem than members of other more numerous species in the system because it alone fulfils a role in the northern rainforests that in other ecosystems is performed by a variety of species. The Southern cassowary (there are three main species, the other two living only in islands to Australia's north) feeds on the fruit of several hundred forest species. It eats the fruit whole and the seeds are deposited on the forest floor in its scats. Its seed dispersal role is vital as, unlike in many other ecosystems, the cassowary is the only seed distributor for more than 70 plants in the northern rainforest and assists more than 80 other plant species by its eating habits.[10] Some plant species, such as the rare forest tree *Ryparosa*, are almost totally reliant on the seeds passing through the cassowary's gut to germinate at all.[11] The Southern cassowary is a keystone species in its environment because without the cassowary the ecosystem in which it lives will deteriorate, the biodiversity of the

10 http://www.wettropics.gov.au/pa/pa_casso.html (accessed June 25, 2009)

11 Bruce Weber and I.E. Woodrow, Cassowary frugivory, seed defleshing and fruit fly infestation influence the transition from seed to seedling in the rare Australian rainforest tree, Ryparosa sp (Achariaceae), Functional Plant Biology, Nov. 1. 31: 505-516.

system will be (possibly, radically) diminished, many individual organisms will be threatened and some species will be eradicated. Its impact on other species therefore gives it also a high conservation value.

Both capacity or potential to affect and actual interconnection are morally relevant factors. But it may not be a simple matter to separate them. The cassowary is a relatively complex organism and its capacity to affect the ecosystem is in part dependent on its complexity, but not entirely. There is something in the way the cassowary interacts with its environment that makes it so important, but in other ecosystems it may be a simpler, or even a more complex, entity that determines the ongoing health of the system. The cassowary is an example of actual effect, although, of course, it must always have had the potential. The actualisation of an organism's potential depends on a range of factors many, if not most, of which are external to the organism.

How an organism is organised internally and how it relates to its external world are clearly important. It may be that there can be degrees of intrinsic value and that more complex organisms have more intrinsic value than less complex organisms. And it may be that on this basis more complex organisms deserve priority over less complex organisms when. But how are we to understand what it means to be complex?

In systems terms, living organisms are thermodynamically open (there is an exchange of energy between themselves and their environment) and organizationally closed (the functional relationships between internal components remains constant) and the organism is able to persist in its being as a result. The complexity of organisms and their interactions with the environment can be put in terms of autopoietic systems. And while autopoiesis may not be the best candidate for the job, a brief account of some of its key elements is still useful for understanding the issue of complexity. Autopoiesis, while thought to be a necessary and sufficient condition for the presence of life by those who support the concept, properly refers only to the *internal* organization of single cells and discrete multicellular organisms. Maturana and Varela, the originators of the concept, regard organisms as having a mechanistic operation and argue that all machines, living or not, have an *organization* and a *structure*. For them, the *organization* of a unity is the 'relations between components that define a composite unity (system) as a composite unity of a particular kind'[12] and the behaviour of any autopoietic system is completely determined by its structure. On closer examination, this structure determinism appears to be a way of understanding causation that draws on something like the notion of causal powers, an idea common among the ancients and medievals which has been given a modern elaboration by Harré and Madden[13]. One of the key elements of Harré and Madden's treatment is the notion of a 'powerful particular', the possession of which allows an entity to initiate change.[14] That is, living things may be seen to have a unique quality (expressed in chapters 1 and 2 in terms of Aristotle's concept of *dunamis*) that allows them to initiate change in themselves and in others. Maturana and Varela's structure determinism can be expressed in terms of the Aristotelian concept of *dunamis* in that every object has within it a passive dynamic principle (*dunamis*) that allows it to be moved or changed in some way when acted upon by a matching active dynamic principle. This matching active dynamic principle may operate from outside the object or, in the case of living things, may also operate from within. Living things may employ this active dynamic principle on other things, on themselves *as if* they were an other, or on themselves *as* themselves.

In the theory of autopoiesis, structural coupling is what allows cells to aggregate and develop into multi-cellular organisms. It is also what allows a cell to interact with its environment, and what allows multi-cellular organisms to interact with other organisms and their environment generally. And, pertinent to the following discussion of von Uexküll's *Umwelt* concept and biosemiosis, it is also claimed to

12 Humberto Maturana and Francisco Varela, Autopoiesis: the organization of the living, in Maturana and Varela, Autopoiesis and Cognition, 1980, p. xix.

13 Rom Harré and E. Madden, Causal Powers, Oxford: Basil Blackwell, 1975.

14 Ibid., p. 50.

be necessary for communication to be possible between organisms. To quote Maturana:

> Two plastic systems [that is systems capable of changing in response to stimulus] become structurally coupled as a result of their sequential interactions when their respective structures undergo sequential changes without loss of identity. Therefore, the structural coupling of two independent structurally plastic unities is a necessary consequence of their interactions, and is greater the more interactions take place. If one of the plastic systems is an organism and the other its medium, the result is ontogenic adaptation of the organism to its medium: the changes of state of the organism correspond to the changes of state of the medium. If the two plastic systems are organisms, the result of the ontogenic structural coupling is a consensual domain, that is, a domain of behaviour in which the structurally determined changes of state of the coupled organisms correspond to each other in interlocked sequences.[15]

One of the consequences of the establishment of so-called 'consensual domains' is that it allows, for those entities capable of sharing in it, a meta-domain of representations. Such a meta-domain is a semantic, or meaning-making, domain that employs a shared sign system to impose a consensual order on the environment of those entities sharing the sign system. For Maturana and Varela, communication for organisms generally, and language for humans particularly, is a function of structural coupling, a process that may act as an indicator of complexity. They suggest that human communication may be a strict analogue of *trophallaxis*, the chemical-based communication in insects such as ants. They give this linguistic trophallaxis the name *linguallaxis* and say that 'language is a permanent biologic possibility in the natural drift of living beings'.[16] Maturana says that structural coupling is basic to and constitutive of what is commonly called 'cognition' and that because any system that is structurally plastic is a

15 Humberto Maturana and Francisco Varela, Autopoiesis: The Organization of the Living, p. 327.

16 Humberto Maturana and Francisco Varela, The Tree of Knowledge: the biological roots of human understanding, Boston: Shambhala, 1992, pp. 211-212.

learning system, structural coupling should be considered the building block from which all human and animal interactional systems are constructed. In essence, the more complex the organism the greater its capacity for structural coupling and the richer its learning. This needs to be understood, however, in the context that cognition in the theory of autopoiesis is a biological phenomenon that describes how an organism processes the raw data it receives from the environment in which it lives and how it makes use of that data to survive and prosper. Maturana and Varela claim that all properties of autopoietic unities are entirely determined by their structure. From a viewpoint that assumes autopoiesis operates, any changes that occur in one system as it is structurally coupled to another are a result of the 'triggering' of events that are completely determined by the structure of the system. That is, the structure of the system determines what changes are possible. The more complex the structural coupling the greater the scale of change that is possible. Maturana and Varela's views on causation and structural coupling are, however, a little puzzling. The idea that organisms are part of a semantic, or meaning-making, domain that employs a shared sign system is, however, worth pursuing and in this matter there are clear overlaps between autopoiesis and von Uexküll's concepts of *Umwelt* and *Innenwelt* in that they both 'build on a self-organized, circular view of the organism and its life world seen as a whole'.[17]

For von Uexküll, organisms respond to perceptual signs (*Merkzeichen*) received from their environment. These perceptual signs then 'trigger specific action impulses or operation signs (*Wirkzeichen*)'[18] within the organism. Von Uexküll's concepts, while overlapping with concepts contained in the theory of autopoiesis, are more useful in understanding complexity and how it relates to value as they incorporate explanations of both the structural complexity of

17 Søren Brier, Foreword: Sebeok's Biosemiotic Contribution to Cybersemiotics, Cybernetics and Human Knowing, Vol. 10, No. 1, 2002, p.5.

18 Paul Bains, Umwelten, Semiotica, 134:1/4, 2001, p. 140.

organisms and the complexity of their relationships to—or communication with—the objective external environment.

As far back as the Stoics there has been awareness that 'animals… communicate with each other by means of signs'.[19] But is this morally significant? The following examines the question of whether an entity's phenomenological creation of a world through its ability to 'interpret' signs is morally significant and whether an organism's structural complexity and the consequent complexity of its sign-use capacity is relevant in determining moral considerability relative to other entities. In short: does a more complex semiotic engagement with the objective world through more complex sensory structures imply greater moral value?

Von Uexküll (1864-1944) begins with the Kantian position that we can not know a thing in itself (*das Ding an Sich*) but can only do so through the intermediary of our senses. He then applies this beyond the human. In effect, he says, like humans, other organisms cannot have unmediated access to things in themselves, but make 'sense' of the world in species-specific ways that depend on the particularities and complexity of their internal structures.

All organisms have an engagement with both an inner and an outer world. The inner world, what he calls the *Innenwelt* is a function of the communication between the components of an organism. The organism and the outer world sensed by it together form an interactive unity, an *Umwelt*, or 'subjective' world produced by the being's own manner of processing signs received from the external or objective world.[20] Von Uexküll described this *Umwelt* as a sort of 'soap bubble' created by the interaction of the internal receptors of each organism with signs received. While plants and probably fungi lack the effectors and receptors to do so,[21] every representative of every protist, bacteria and animal species has its own *Umwelt* that it is always actively

19 Thomas Sebeok, The Sign Science and the Life Science, Applied Semiotics, 3:6/7 1999, p. 93.

20 In Husserl's terms, for humans this is the *Lebenswelt.*

21 Martin Krampen, No plant – no breath, Semiotica, 134:1/4, 2001, p. 419.

creating as it reacts distinctively to signals from the physical world. Their internal structures—at whatever level of complexity—allow each organism not merely to receive signals from the outer world, but to react to these signals (or signs) in terms of their own needs. In receiving and responding to signals from the external (objective) world, organisms have the three basic elements of a semiotic relationship with that external world: *object, interpreter* and *sign.*[22] It is this semiotic relationship that allows the organism to create an 'understanding' of the world it inhabits. In doing so it creates for itself a unique 'semiotic niche'[23] the nature of which is defined by its own structure. Simpler organisms (von Uexküll began with the tick[24]) have fewer receptor cells than more complex organisms and so are less able to respond to a wider range of perceptual signs (*Merkzeichen*) than more complex organisms and so have a less complex sense world (*Merkwelt*). The more complex an organism, the more receptors and effectors it has and the more signs it is capable of receiving and interpreting. In essence, more complex organisms occupy a larger semiotic niche than simpler organisms. As a result, more complex organisms also have a greater possible sphere of influence.

More complex organisms have a greater potential to interact with their environments because they have more receptors through which to receive signals and greater internal capacity to interpret these signals. Higher animals, such as humans, also use the signs received to create an internal representation, a counter-world or *Gegenwelt*, 'which arises when a nervous system intersperses neural excitation

22 Marcello Barbieri, Has Biosemiotics Come of Age? in Marcello Barbieri, ed. Introduction to Biosemiotics, Dordrecht: Springer, 2007. Uexküll is generally regarded as a 'cryptosemiotician' in that he did not use the notion of semiotics himself, but developed terms and concepts that were picked up later as important precursors to the biosemiotics identified by Thomas Sebeok.

23 Hoffmeyer, 1996, p.140, cited by Morten Tønnessen, Umwelt Ethics, Sign Systems Studies, 31.1, 2003.

24 Jakob Von Uexküll, Umwelt und Innenwelt der Tiere, Berlin: J. Springer, 1909.

and motor systems between the external world and behaviour'.[25] Maurice Merleau-Ponty's account of von Uexküll's theory explains succinctly the relationship between the various elements.

> 'The higher animal … constructs an *Umwelt* that has a *Gegenwelt,* a rejoinder in its nervous system. In his work in 1934, Uexküll specifies this notion of *Gegenwelt.* He distinguishes the *Welt* (the objective world), the *Umwelt* (the milieu tailored to the animal), and the *Gegenwelt* , which is made up of two systems: the *Merkwelt* and the *Wirkwelt.* The *Merkwelt* depends on the way in which the sense organs are made. These realize a classification of stimuli according to a disposition proper to the animal. The *Merkwelt* is a 'grill' interposed between the animal and the world. In order to determine the world of the animal, we have yet to make the *Wirkwelt* intervene; that is, the reactions of the animal in the milieu, the melodies of impulses. In order to grasp the world of the animal, we must not only make perceptions intervene, but also behaviours, because these deposit a surplus of signification on the surface of objects. The Wirkwelt displaces the Merkwelt[26]

Von Uexküll explains these latter two terms with the example of a bell. '… I put a bell on the table and let is sound… the bell appears in two roles, it has an *effectual* side (*Wirkseite*) that affects our senses and a *perceptual* side (*Merkseite*) that is impressed on the bell by our perceptual centres in the brain and which consist of human perceptual cues…'[27] The organism receives perceptual signs from its environment and these trigger action impulses in what von Uexküll called a functional cycle (*Funktionskreis*). Functional cycles occur in sequences. The first such cycle described by von Uexküll was that of the tick in its interaction with a mammal. The tick is blind and deaf. It sits on grass or foliage and waits for a mammal to pass by. In the first

25 Carolyn Ristau, ed. Cognitive Ethology:The Minds of Animals, Essays in Honor of Donald R. Griffin, Mahwah, N.J: Lawrence Erlbaum Associates, 1991, p.81.

26 Maurice, Merleau-Ponty, *Nature: Course Notes from the College de France*, Dominique Seglard, (Compiler), Transl. Robert Vallier, Evanston: Northwestern University Press, 2003, p. 172.

27 Jakob Von Uexküll, The New Concept of Umwelt: A Link Between Science and the Humanities, Semiotica, Issue 134, 2001, p. 111.

stage of the functional cycle butyric acid from the sweat of mammals triggers perceptual signs in the tick that induce the tick to let go and (with luck) fall onto the passing mammal. In the second phase, the mammal hair triggers a perceptual sign to search for a warm bare patch of skin (in humans these would typically be areas such as armpits, behind the ears or in the groin). In the third phase, the temperature of the mammal skin in these bare areas triggers perceptual signs in the tick that initiate the piercing process, where the tick injects its proboscis into the skin and withdraws the blood it feeds on. A key point for von Uexküll is that of the hundreds of stimuli coming from the mammal's body, only three bear the receptor cues for the tick (butyric acid, hair and temperature). For von Uexküll

> 'living organisms respond to perceptual signs (*Merkzeichen*) or 'meaning' (*Bedeutung*), not to causal impulses. Physical, chemical, or thermal changes to the receptor organs are interpreted as signs of the (not yet perceptible) 'perceptual cues' of an object, as a counterpart for a specific behavior. Uexküll argues that the 'subject' (tick) and 'object' (mammal) dovetail into one another and constitute a systematic whole or functional cycle. The organism or interpreter receives signs from its environment and these perceptual signs trigger specific action impulses or operation signs (*Wirkzeichen*). The whole cycle is a process not made of static objects but of sign relations—a semiosis.'[28]

The *Umwelt* consists of sign relationships. That is, as an interpreter of signs from its (objective) environment the organism is always actively creating its *Umwelt* in a process determined by its structure or design, its activities and its needs. Throughout the functional cycle there is a counterpoint or contrapuntal relation between the organism as a meaning-maker and the perceptual cues of the object. This is described as a sort of nature 'music'[29] with its own harmonies.

28 Bains, Umwelten, p.140.
29 Ibid., p.141.

In the *Umwelt* theory, all organisms respond to signs but *Umwelten* can only be attributed to the protists, bacteria and animals. Plants, and probably fungi, have *Wohnhüllen.*[30]

> 'in which the objects of *Umwelten* are replaced by meaning-factors which are still a category of individual phenomenal worlds. [However], while only *Umwelt*-carriers take part in functional cycles, plants and fungi, as well, partake in contrapuntal relations i.e. subject-object-relations characterized by a mutual correspondence between the two entities.[31]

The known phenomenal world consists of *Umwelten* and *Wohnhüllen* and the interconnectedness of these is what we commonly call nature. Plants do not have *Umwelten* but they share contrapuntal relations with beings that do. The capacity or potential of an organism to interact with its environment through contrapuntal relations defines an 'ontological niche'.[32] And those with a more complex structure of receptors and effectors occupy a more complex ontological niche. Organisms of all sorts have an ontological niche value as well as a 'semiotic' and an ecological niche value. But more complex organisms occupy a richer phenomenal 'space' or 'area' than less complex organisms. They may or may not have a greater intrinsic value as a result, but the 'niche' occupied by more complex organisms is larger. With this larger niche comes a greater richness of meaning-making, a greater freedom for semiosis, but also a greater capacity to affect the phenomenal lives of other living things. The size of the aggregate niche is relevant. Other things being equal, removal of a more complex entity from its physical environment has a greater effect—through the absence of contrapuntal relations—than removal of less complex entities. But is this enough?

Let's go back to some basics. While personhood and rationality may be sufficient for moral considerability, there is wide agreement

30 A rough and ready translation of this might be 'dwelling envelope' – a sort of 'space' in which the entity lives.

31 Morten Tønnessen,Umwelt ethics, Sign Systems Studies, 31.1, 2003, p. 288.

32 Ibid.

that they are not necessary. And mere membership of this species or that species does not explain the moral claims that might be made either by or on behalf of an individual, so what else might? Sentience is often cited as a sufficient condition for moral considerability and, crucially for the discussion here, it may be necessary.[33] Sentience, however, is not an unproblematic notion, but if we take its meaning at the most general level, say Sellars' definition of sentience as mere sensations, and as something to which the phenomenal belongs,[34] then all entities with *Umwelten* are sentient and *prima facie* morally considerable. But we still need a way to discern moral difference.

The above shows that (certain) organisms are active subjects which create meaning for themselves as they encounter the external world. This ability to create meaning (through biosemiosis) is a centrally important element of being alive. Complexity is important in creating meaning because it contributes to the richness of the biosemiotic world and to the richness of ontological diversity. Hoffmeyer[35] suggests that this richness, expressed as semiotic freedom, is a parameter that might be used for distinguishing between moral subjects. From the above we know that as a species dies out the richness of the biosemiotic world is diminished and an ontological and semiotic niche disappears—never to reappear. Something of value—at least to those entities that shared the *Umwelt* of the now-departed—is gone. We also know that living organisms with an *Umwelt* share a common phenomenal life with other organisms and with greater semiotic richness comes greater sharing. That is, more complex organisms receive and process more signs from the objective world

33 Lori Gruen, The Moral Status of Animals, The Stanford Encyclopedia of Philosophy (Fall 2010 Edition), Edward N. Zalta (ed.), URL = <http://plato.stanford.edu/archives/fall2010/entries/moral-animal/>.

34 Wilfred Sellars, Empiricism and the Philosophy of Mind (1956), Cambridge: Harvard University Press, 1997.

35 Jesper Hoffmeyer, Biosemiotics and Ethics, Nina Witoszek & Elizabeth Gulbrandsen, eds. Culture and Environment: Interdisciplinary Approaches, Oslo: Centre for Development and the Environment, 1993.

and consequently have a richer *Umwelt* than simpler organisms. Some also have a *Gegenwelt*.

We (humans) share a common life with animals, and as Stephen R.L. Clark would say, they have a pull on us: they make a claim. And the greater the semiotic richness, the greater the pull or claim, the greater the call, the greater the resulting obligation. We can start to understand this by looking at what happens when we meet another human. When we face another 'face' we experience the greatest pull. We enter into a shared world which we recognise will be richer than any we might experience with another animal. In recognizing the human other we make a distinction that is, at least in part, based on the perceived semiotic richness of the *Umwelt* we have begun to share, and on facing another human we mark her as having greater value than non-humans we might encounter. If we were to extend this notion a little, those with semiotically rich *Umwelten* might be considered what Tom Regan calls 'subjects of a life' and—in a manner analogous to our obligations to humans—be due moral consideration. As semiotic richness declines the pull or demand or claim or call on us diminishes: animals with less rich *Umwelten* pull us less than animals with richer *Umwelten*. In this extension our obligations to the higher (non-human) animals will be qualitatively different from obligations to humans, and obligations to animals and other living things will diminish with diminution of semiotic richness. In this it is not that a Lévinasian 'face' confronts us when we encounter non-humans (although with the higher primates at least, there is something very close to this), but we might claim with some surety that what happens is an analogue of this reaction. This idea is explored further in the next section.

Humans may be a complex (even if the most complex) organism and as such may have greater claim on us than, say an amoeba or a snail. But human claims, no matter how great, must be weighed against the value of the actual interconnections humans make with their environment. The value that humans have to the environments they encounter is often negative and would appear to be linked to their complexity of organization as this gives them greater capacity to act in a way that affects the interests of the other organisms around them. Even though humans may have great value because they are highly-complex organisms their ecological niche value also needs to be taken into account (as it needs to be for all other organisms). Humans, unfortunately, are more likely to behave in the manner of what might be called a '*millstone*' species—a species in whose presence an ecosystem is likely to collapse, or at least significantly diminish. However, despite their frequently negative ecological niche value, humans occupy a unique, and rich, semiotic and ontological niche. They are agents with a huge capacity to affect their environment, but they are also *valuers par excellence* who both recognise and value *value* itself and are able consciously to act to protect it.

Responsibility

Humans can choose to protect value, but why should humans, as moral agents, consider the wellbeing of others, whether those others are fellow humans or individual members of other species?

For Emmanuel Lévinas, because we are subjects, we necessarily have responsibility for the Other (in the abstract), and hence, contingently, for actual others. This responsibility for the Other 'arrives before the Other herself had the time to demand anything'. This is a 'responsibility before commitments have been entered into, a responsibility that is the *a priori* measure of all commitments, rather than being *a posteriori* measured by them.'[36] From this perspective we need to acknowledge an *a priori* responsibility *for* the generalised or abstract Other which is the basis of a practical responsibility for, and obligation to, the concrete other in front of us. At the concrete level, recognising a 'face' puts us under an obligation where the cognizing

36 Zygmunt Bauman, Postmodern Ethics, Oxford: Blackwell, 1993, p. 85.

subject (I) is 'responsible for the Other without waiting for reciprocity'[37] and where 'when the face is encountered for what it is, …the inexhaustibility of what it signifies—the Other—is welcomed as a Good-in-itself for whom I am responsible'. But if we are to use Lévinas' theory to apply beyond human-human interaction we must do so with caution, not least because Lévinas' use of the I-Other distinction applies as an abstraction, and not primarily at the concrete level. His intention is to paint ethics as 'first philosophy'. Lévinas's Other is not just another You and the relationship is not just one between an 'I' and a 'You'.[38] This Other has meaning for itself quite independently of any value I bring to the encounter and my obligation to it is prior to and greater than any obligation I might owe to myself. This priority of obligation is because it (the concrete other) faces me immediately and directly before I face myself (I only face myself in reflection) and announces a need to which I have an obligation to respond. This responsibility to the Other is the 'a priori measure of all commitments rather than being a posteriori measured by them'. We should ask, however, what entities count in this. That is, what constitutes a 'face' that, once recognised, imposes an obligation of responsibility on us? It is clear that we have obligations to humans, but what of other beings? And what might the obligation, the responsibility, be?

Hans Jonas notes that the recognition of values in nature does not bind those who recognize the values to act in any *prescribed* fashion, but maintains that despite this, 'by showing the immanence of purpose in nature...the decisive battle for ethical theory has already been won'.[39] That is, there is an imperative to do *something* even if it is not clear just what is to be done. His argument here relies on the Aristotelian proposition that purposiveness is an *ontological* characteristic of an entity and that the mere capacity to be purposive is a good-in-itself. He maintains that '...it follows already analytically

37 Lévinas, Ethics and Infinity.

38 Vogel, Jewish Philosophies after Heidegger, p.128.

39 Jonas, Responsibility, p. 78.

from the *formal* concept of the good-in-itself that, whenever this first, self-validating good happens, in any of its individuations, to come under the custody of a will, it addresses an 'ought' to this will'.[40] The good-in-itself is an end in itself that imposes a duty of care on any moral agent. In the first instance this is an unspecified duty which needs subsequently to be made concrete. To make this responsibility concrete moral agents should ask: What do you need? And not merely give what they think is needed.[41]

In Aristotelian metaphysics living organisms are ends-in-themselves with an immanent claim to the realization of their own particular *telos*, which entails them being alive and striving to flourish—whatever flourishing is for them. As such they are also fit subjects for a moral imperative such as that expressed in Kant's formula of the end in itself. That living things have a claim to be the realization of a good is consistent with the Aristotelian position that there is purposiveness in nature itself. That is, from an Aristotelian perspective it is meaningful to speak of immanent, *even if involuntary and unconscious*, purposiveness in living things. And it follows directly from the immanent purposiveness of ends-in-themselves that there is value in the world which needs, itself, to be valued. As Jonas says, 'Nature harbors values because it harbors ends and is thus anything but value-free'.[42]

But a generalised obligation to ends-in-themselves is not sufficient to guide us, as moral agents, adequately in our dealings with the world. What then, might help us to differentiate, to tailor our behaviour in a consistent and justifiable way? In *The Imperative of Responsibility*, Jonas sets out a theory based on the axiom that responsibility is the correlate of power, that is, because one living thing (in his case a human) has the power to affect the wellbeing of

40 Ibid., p. 80.

41 This would seem to be an argument for taking heed of good science. And it is, but only to the extent that the science itself is well-founded. That is an important matter, but one outside the scope of this present work.

42 Jonas, Responsibility, p. 78.

another a relationship of responsibility is established. His paradigm case of responsibility is that which a parent owes to its offspring. He argues that the newborn in its facticity imposes an ought-to-do on the parents: simply because the newborn exists, and its parents caused it to be brought into being, the parents ought to care for it.

The idea of responsibility is complex, but we should at least note that for each of us there is a *de facto* role responsibility on the basis of which we will justifiably treat apparently similar others in a differential manner. For example, a parent will likely (and defensibly) favour his or her own children over others in certain situations; or a lawyer is required to favour his or her client over others. This is not to deny the existence of an undifferentiated responsibility to unspecified others—based on an *a priori* responsibility to the Other, it does, however, acknowledge that recognising responsibility is the start of a process. We are beings who are in a variety of relationships and those relationships need to be taken into account when determining moral responsibility.

Responsibility is a complex notion and there are many ways in which we may think of it, but at the least there is both a retrospective and a future-looking sense in which we can understand it. In acknowledging retrospective responsibility for an event, Y, I am agreeing that my actions in some way contributed to event Y coming about whether or not I intended event Y to occur—although intention, accident, carelessness and inadvertence each bring a different moral flavour to the events and my actions. In acknowledging a future-looking sense of responsibility I am acknowledging an obligation to act in certain ways in the future. For example: with her partner, a woman is responsible (in the retrospective sense) for her children coming into the world. She and her partner also have a responsibility into the future to look after the children.

There is also another homophonous sense to take into account: in acknowledging that there is a situation in need of a remedy—whether my actions contributed to the situation or not—I need also to determine whether I have the *ability* or *capacity* to respond. In Warren Reich's terms I need also to have 'response-ability'.[43] As a human I may have obligations to care for particular others, but in practice I may not always have the ability, capacity or authority to act. For example, an Australian living in Perth may recognise a need to assist Florida alligators or Yangtze dolphins but will of necessity be constrained as to what he or she can do. There are practical limits. There are also practical limits to our response-ability determined by the roles we take on. For example, a committed environmental activist may be limited by scarcity of resources, or a lawyer serving a client wishing to bulldoze a wetland is limited in what she can do because of the role she has undertaken and the consequent fiduciary duty to her client.

Response-ability may also be seen to figure in the distinction between substantive and procedural issues in ethics. We may all agree that there is a problem (that there is a *substantive* issue) but not be able to agree on whose responsibility it is to act (a *procedural* issue). A school teacher observing a child being bullied in the playground has a responsibility to act based on her role as a teacher. In procedural terms she has a responsibility, and the power, to act. The same teacher observing a similar event between two different children on a Sunday afternoon at a football match does not have the same procedural responsibility. She may feel obliged to intervene as a citizen, but in this role she does not have any special power, authority or duty to act. Her response-ability outside of the school is limited by the procedural fact that she has no special charge, over and above that of other citizens, to intervene.

Responsibility presupposes the presence of agency. That is, there are agents who have causal powers through which their acting can have an effect on the world and who are both able to act and able to foresee to some extent the consequence of their actions. For example,

43 Warren Reich, The Duty of Care: Implications for the Profession, paper presented at the Australasian Association for Professional and Applied Ethics 13th annual conference, Sydney, June, 2006

I am in one sense responsible for event 'X' if X was a result of my actions, whether the result was intended, accidental or inadvertent. One has a (retrospective) responsibility *for* one's own deeds, in the way that, say, humans are responsible for destroying the habitats of other animals or in the way that I am responsible for having broken a window if I hit it with a hammer. This sense of responsibility is tied to the causal powers of an agent: 'I am responsible' in this first sense might be replaced by 'I am responsible—I performed an action that led to it happening; I did it'. In contributing to X I may also be thought to have behaved responsibly if the intended or actual result is justifiable or to have behaved *ir*responsibly if I should have foreseen that X would have a deleterious effect (on myself or others). If I 'take responsibility' for event X happening I acknowledge that my actions were a causal factor in event X occurring—but I may also be acknowledging an obligation (forward-looking responsibility) to remedy any wrongs that result from event X. So, a second sense might be seen as (forward-looking) responsibility *for* particular objects or particular beings, a sense that carries with it an element of *care*. It is in this second sense that parents are responsible for the welfare of their children, i.e. there is an onus on parents to ensure the continuing wellbeing of their children. Or, if I borrow a friend's car, I am responsible *for* it in this second sense in that I must try to ensure that the car is not stolen or damaged in any way. 'I am responsible' in this second sense might be expanded to 'I am responsible for the object—I will ensure that it is cared for'.

The first sense of responsibility *for* above, Jonas calls *formal* and the second *substantive*. Jonas' 'substantive' sense of responsibility is not the same as that in the substantive/procedural distinction noted earlier but aligns with the future-looking sense just discussed and it is this sense that carries a moral imperative. However, it is significant in the above passage that a responsible moral agent will 'ensure' that the subject is cared for rather than necessarily care for it him or herself, as this points to a difference between care and responsibility—a difference explored shortly. It is worth noting that responsibility is an abstract noun the full meaning of which also includes a sense of behaving responsibly. In the example above if I hit the window with a hammer I may or may not be behaving responsibly. If I hit the window out of whim or malice I could be said to have behaved irresponsibly. If I break the window to enter a house to save a child in danger I would have behaved responsibly.

'Formal' responsibility in Jonas' terminology refers to what has already been done, whereas 'substantive' responsibility concerns 'the forward determination of what is to be done' and the responsibility is primarily to 'the *matter* that has a claim on my acting'.[44] Here we may infer that for a matter to have a claim on my acting, the 'procedural' issues and the attendant 'response-ability' have been sorted out. In the case of substantive (forward-looking) responsibility *for* others, what I am responsible for is outside myself, but within the effective range of my power. And in the case of living things, what I am responsible *for* is a 'good-in-itself' that makes an appeal to me and 'confronts my will and demands to be heard'.[45] In Jonas' words, 'first comes the 'ought-to-be' of the object, second the ought-to-do of the subject who, in virtue of his power, is called to its care'.[46] In effect, *can* implies *ought*. If it is in a moral agent's power to help, she or he ought to help.

In trying to clarify his concept of substantive responsibility, Jonas turns to an examination of *ir*responsible action. He says that the gambler who puts his own fortune at risk acts recklessly. When it is another's money put at risk the gambler's action might be criminal, but when the gambler who risks his fortune has a family that depends on him, he acts irresponsibly. Likewise, a reckless driver who endangers his passengers' lives also acts irresponsibly. In each of these cases of *ir*responsibility 'there exists a definable, *non*reciprocal *relation* of responsibility'.[47] Under such a nonreciprocal relationship, the mere fact of my control or influence over the fate of others places me under an obligation to act in certain ways. This obligation is

44 Jonas, Responsibility, p.92.
45 Ibid., p. 85.
46 Ibid., p. 93.
47 Ibid.

independent of whether the others—human or non-human—are aware of my capacity to influence their fate. In cases where the other is aware of the capacity to influence and is aware of my responsibility then acts of irresponsibility are also breaches of trust, where trust is 'reliance on others' competence and willingness to look after, rather than harm, things one cares about and which are entrusted to their care'.[48] But trust itself is a complex and problematic notion that will have to be left unaddressed here.

Human-nature Dualism

Jonas extends the object of the nonreciprocal relationship beyond the human sphere to take in a sense of responsibility for the natural world. In doing so he is in large part mounting a counter-attack on the nihilism of Heidegger which posits a separation between humans and nature and presents a nature that has no ends, is subordinate to the ontological status of humans, has no value and which is incapable of caring. Heidegger's nihilism creates an ethical vacuum with respect to human attitudes to and actions toward nature. If one were to take a position on the natural environment based on Heidegger then 'nature'—as 'phūsis', that which emerges[49]—would be something to be wrought for human ends and would have, at best, instrumental value. Jonas recognises there is more than mere instrumental value in nature; that nature has value in and of itself and that humans have a responsibility to take this value into account. More than this, he says that nature *cares* because 'living nature, from which our own caring selves emerge, and on which we depend, is essentially good, is worth

48 Annette Baier, Trust and antitrust, Ethics, Vol. 96, No. 2. Jan. 1986, p. 260.

49 Martin Heidegger, An Introduction to Metaphysics, Trans. R. Manheim, Newhaven: Yale University Press, 1959, p.71.

being cared for and even cares that we care for her so that we, her most magnificent creation, can continue to be.'[50]

Humans may or may not be nature's most magnificent creation, but they are not the sole locus of value in nature. However, the mere existence of value in nature docs not, of itself provide a basis for a forward-looking responsibility for naturc. Value in nature is objective value, a good-in-itself. In Jonas' view, this good-in-itself 'addresses an 'ought' whenever it comes under the custody of a will' and the 'will must experience itself as responding to a transcendent summons in order for the moral sentiment' to be more than impulse. It is a 'command of Being that arouses our *conscience* and calls us to take responsibility.' [51]

However, although nature may be 'good' and may 'care', Jonas claims that only humans may exercise responsibility and that the first imperative is to ensure the persistence of humanity. Jonas's claim is itself based on a claim that humans alone can have power *and* reason and that power conjoined with reason carries responsibility with it. The question of what may exercise responsibility, however, is problematic as, *prima facie*, the behaviour toward their offspring by the higher primates and the cetaceans, certainly, and birds and some fish, probably, would seem to give lie to the claim that only humans have the capacity to exercise responsibility. It would have to be an arbitrarily narrowed definition of responsibility for it not to recognise that the nurturing behaviours of some at least of the non-human mammals constitute instances of being responsible for, of *taking* responsibility for offspring. A distinctive difference between human responsibility and that of other animals may lie in the notion of *behaving responsibly* and the social meaning that accompanies it. For human acts of *ir*responsibility to offspring there is some social

50 Lawrence Vogel, The Outcry of Mute Things: Hans Jonas's Imperative of Responsibility, David Macauley ed., Minding Nature: the Philosophers of Ecology, New York: The Guilford Press, 1996, p.171.

51 Ibid., p. 175.

sanction or penalty, whereas social sanction or penalty is not obvious in non-humans.

Jonas does not limit the *object* of exercised power and reason to humans, but extends the imperative of responsibility to non-human nature. However, the extension of responsibility is still limited and grounded in a form of anthropocentrism that comes down to the claim that because man has the might he has the right.

> '...in the choice between man and nature as the struggle for existence poses it time and again, man indeed comes first, and extrahuman nature, even with its own rights acknowledged, must give way to his superior right...the mere fact of superior power legitimates its use.'[52]

He may have moved away from Heidegger to acknowledge value *in* nature, but he has deliberately retained an anthropocentric bias. One of the consequences of this anthropocentrism is that, even when he turns his attention to an ethical relationship to the environment he focuses on human interests, and specifically on the benefits to humans of ensuring that the natural environment is protected and nurtured. Lawrence Vogel—one of the best-informed commentators on Jonas—believes, however, that 'Jonas's metaphysics undercuts the very distinction between anthropocentrism and nonanthropocentrism...'[53] in such a way that both positions can be maintained. In Vogel's interpretation of Jonas, living nature is a good-in-itself, is comprised of ends-in-themselves, and *commands* us to show regard for it. The special value of humans is that of moral agents to whom the imperative of responsibility is addressed and who can choose to act in the interests of others. The anthropocentric bias is explained through the claim that the first duty of humans is to maintain moral agency itself.

But, the very power of human agency in general, conjoined with reason, has disturbed the balance of nature in fundamental ways.

52 Jonas, Responsibility, p. 137.
53 Vogel, The outcry of mute things, p. 180.

Human agency and the responsibility attendant on it has, for Jonas, been well understood in the intrahuman sphere over many centuries but there is much yet to be understood about the 'novel expansion of responsibility to the condition of the biosphere and the future survival of mankind...'[54] In the current awareness of ecological crisis there is a 'heuristic of fear' in which we speculate on what might happen to indeterminate others, including those who are not yet born. That is, fear of an unpleasant outcome is used to motivate action. There is fear in the original speculation and any active responsibility begins with the question: 'What will happen to *her* [the unspecified individual] if *I* do not take care of her? The darker the answer, the more clearly the responsibility appears.'[55]

Jonas' expansion of responsibility to include responsibility for improving the condition of the biosphere bears closer examination. However, before embarking on that examination, we should note that if (substantive) responsibility is to be taken for improving the biosphere (for whatever reason) then any action has to be taken in concert—and after negotiation—with other members of society. This is so partly because the job is simply too big for any individual alone, but also because any group effort must take account of the possibility that moral agents might be faced with conflicting responsibilities. Consideration of this social context of responsibility cannot be undertaken without also considering what Jonas calls *political responsibility,* but nor can it be undertaken without moral agents, individually and then collectively, *accepting* responsibility, accepting an obligation to protect and further the interests of non-human nature into the future.

54 Ibid., p. 138.

55 Jean Greisch, Otherness: an Ethical Category? Guillaume de Stexhe and Johan Verstraeten, eds. Matters of Breath: Foundations for Professional Ethics, Leuven: Peeters, 2000, pp. 193-194.

Forms of Substantive Responsibility

For Jonas, there are three forms of substantive (forward-looking) responsibility: natural, contractual and political. The relationship of responsibility of parent to child is an example of *natural* responsibility, the original form of responsibility from which the two other forms 'ultimately derive their more or less contingent validity'.[56] Contractual responsibility is an institutional responsibility, such as an employee has to his or her employer (and vice versa). Natural responsibility has a stronger moral status than contractual responsibility. The third form of responsibility is political responsibility, the paradigm of which is the statesman who seeks power in order to gain responsibility—and supreme power for supreme responsibility. Jonas' account of political responsibility has the possibly unfortunate effect of validating a 'green' totalitarianism. But more on that later.

The objects of *natural* responsibility are living organisms, with humans taking a position of priority on the grounds that they are probably the only entities that can exercise responsibility. For Jonas, the objects of *political* responsibility are humans alone, although it is not clear from his argument why this should be so.

Let us hold off on who or what might exercise *political* responsibility and return to what can be considered to be objects of responsibility and the question of whether they should be different for natural responsibility than for political. For Jonas, 'the referent of responsibility—the field of its 'causes'—is life, actual or potential and above all human.'[57] It is 'above all human' and greatest consideration should be given to humans because they alone can exercise responsibility. Also, in Jonas' scheme, the highest duty of responsibility is to ensure that responsibility itself is not stifled. Jonas

56 Ibid., p. 95.
57 Ibid., p. 101.

argues that humans should have priority because the possibility of (forward looking) responsibility itself is tied to the existence of humans: only humans can take responsibility. This claim is, however, a long way from being beyond dispute, as debate over the possibility of moral agency in the great apes, for example, indicates. Research into the cognitive abilities of the higher primates, noted in Chapter 4, suggests that humans are not alone in being able to exercise self-reflexive rationality and nor are they alone in demonstrating care and concern for others. So, even though interpretation of the evidence on the higher primates is contentious, it would seem unwise, if not hasty, to limit the exercise of responsibility to humans and any attempt to do so should be considered arbitrary. Pets and guide dogs, for example, do seem to care for the humans close to them; and dogs have frequently been recorded as nurturing kittens, lambs and the like. Also, if the paradigm case of responsibility is that which a parent owes to its offspring and if the newborn in its facticity imposes an ought-to-do on the parents, why would not the facticity of a newborn mammal of any sort impose an ought-to-do on its parents?

Responsibility suggests itself as an important, if not vital, element of a workable ethic and we need at least to *accept* responsibility for the degradation of the environment and *take* responsibility for the future welfare of the natural world. But, again, what are to be the objects of responsibility for agents who take responsibility for others? An answer to this can best be ascertained by truncating the above quote from Jonas so that 'the referent of responsibility...is life, actual or potential'.

The key element of an ethic based on responsibility is that an agent is placed under a nonreciprocal obligation to an other if that other has an intrinsic or immanent claim to be regarded as morally considerable. A living organism has a good of its own that it strives to realize, that is, it has an immanent 'ought-to-be' that in effect makes a *claim* on an agent and places the agent under an 'ought-to-do' obligation. The end of this ought-to-do is the wellbeing of the organism. To re-visit the example from earlier, the newborn baby has an immanent ought-to-be that places the nurturing parents under a nonreciprocal relationship of responsibility: the parents 'ought to do' that which ensures the baby survives and flourishes. The baby 'claims' the parents in the sense that the parents are placed under an obligation to ensure that the needs of the newborn are met and that it flourishes.[58] The newborn, as yet, has no obligation toward its parents or any others. In the absence of its parents, the newborn has a valid claim on any moral agent who can help: the fact that it *is* and that it strives for realization of its immanent good places a demand on any agent who can help it realize its good. But the claim that the newborn makes on moral agents should not be restricted to newborn humans. Surely the newborn puppy or kitten or wombat or whale also addresses an ought to the world around. But, what of the less cuddly species—the cockroaches, the rats, even bacteria? If they too address an ought to the world, moral agents have a *prima facie* obligation to take responsibility for them. This position, which is consistent with a Buddhist or Jain ethic, may be resisted by some moral agents but recognising that a being is morally considerable is not to say that it has an *equal* claim on our consideration as that of other individuals, of whatever species.

We operate within a hierarchy of competing claims in which more complex entities have a greater capacity to affect others than do less complex entities and in which the more complex an organism, *prima facie*, the greater its claim on moral agents. If we accept the idea that a good-in-itself addresses a claim to the world around it then we have a responsibility to take into account the good-in-itself of simple entities, even if they are slimy or verminous. But taking into account such entities should not, all other things being equal, be to the detriment of entities of greater complexity or with greater value to their ecosystems. If a thoroughly unlovable creature low in the order of complexity actually has a very high ecosystemic value, our responsibility for the maintenance of its wellbeing will—for very

58 There are clear similarities between the claim made here and Feinberg's notion of a claim right.

pragmatic reasons—be greater than for others. It has value in itself, to be sure, but its ecosystemic value, a measure of the instrumental value it has to others in its ecosystem, needs also to be taken into account. The value as an end-in-itself and a good-in-itself of each organism in an ecosystem can be protected by giving preferential treatment to organisms with high ecosystemic value because in protecting those with high ecosystemic value the system itself is protected and all elements of the system are thereby protected.

So, what should be done in cases of conflicting demands for help? Semiotic and ontological niche value may offer a way forward. Bio-semiotic niche value and ontological niche value contribute to the total value of a species. And if one were able to establish some sort of calculus of value the total of these values might be good enough reason to value one species over another where ecosystemic niche value alone is not sufficient to differentiate (in value terms) species or individuals. Moreover, those species (notably humans) that can value *value* itself and can act consciously to protect it would have an additional claim to differential treatment.

However, it is still the case that if humans are part of nature and not apart from nature then an endangered keystone species may still have a stronger claim to be the object of responsible action than any given particular human. A keystone species is, by definition, very well interconnected with its environment and actually (not merely potentially) affects many other individuals of different species. It will have a greater claim on a moral agent than does a member of any other species.

The subject of the 'ought' by which moral agents find themselves bound cannot be restricted to members of any one species and no one species has, *a priori*, an overriding right to greatest consideration. Take, for example, any of the baleen whales that have, until recently at least, faced imminent extinction. Because they are major consumers of plankton they have a very high niche value in their environments, to the point that they should be considered keystone species. Each whale is also a store of a unique species value (ontological value) and a store of bio-semiotic complexity both of which would be lost if the species were to die out. It is plausible that the whales (as a species) should be protected even if it means that a number of humans have to die for this to happen. It is not a calculus anyone could conduct with any joy, but it is a calculus that should not be avoided.

On the basis that *individual* living organisms are primary substances in an Aristotelian ontology, and that they are entities with an immanent *telos* and which strive to realize their immanent good, individual living organisms should be considered the primary objects of responsibility. However, each is also a store of species value, bio-semiotic and ecological niche value.

Should we follow Jonas and say that one can only be responsible for what is changeable or perishable[59], and that 'the object of *responsibility* is emphatically the perishable *qua* perishable'?[60] Presumably, that which is not perishable does not require a moral agent to take substantive responsibility for it because the moral agent can have no influence on it. The category of what is perishable obviously takes in all living things, but unless a very narrow definition is applied it also takes in other things as well. A work by Titian or Rembrandt or Klee, for example, is perishable. However, any responsibility that one has toward these, and other artefacts, is a responsibility contracted between a citizen and a state, or between a member of society and the maintenance of that society's culture. The responsibility for a Titian or a Rembrandt or a Klee is different from the responsibility for a living thing. In the case of artworks, our responsibility is not to the artwork itself, but to the investment that society makes in that artwork. As members of a society we need to value the relations which help make us a society. One of those relations rests in the value we attach to artworks. This value is, however, not a value intrinsic to the work of art itself, but is a value assigned to it by the society in which the artwork is situated. The artwork does not have a 'good-in-itself'—at least not in the same

59 Jonas, Responsibility, p. 125.
60 Ibid., p. 87.

sense that a living thing does, because the artwork is not self-maintaining and it does not have an *interest* in its own maintenance.

A moral agent has an imperative to take responsibility for the well-being of those things that have an interest in the persistence of their being: that is, to individual things—one by one—that have a good of their own, i.e. to what Taylor calls teleological centres of activity, to what Mathews calls self-realizing *conatus*-possessing entities or to what Maturana and Varela describe as autopoietic unities. The obligation to take responsibility for, or to behave responsibly toward, things with a good of their own should be tempered with the good that an entity has for others through both its capacity for, and its actual, interconnectedness (as expressed in ecological niche and bio-semiotic terms). That is, the intrinsic good of a living thing needs to be supplemented with other of the living thing's values. But in the absence of clear evidence that one particular organism has a greater niche value than another, more complex autonomous entities should be given priority over less complex entities because their greater autonomy and complexity gives them a greater total value.

In the case of mammals and other complex animals, the issue of individuality is fairly straightforward, but what of groups of insects such as bees and ants with highly-developed social structures? Should we consider the basic ontological unit to be the individual insect, or the colony? If we opt for the latter on the grounds that it might constitute some sort of 'superorganism', along the lines set out by E.O. Wilson in *Biophilia*, then this obviously has a higher level of complexity than any individual. The niche value of an ant colony is also demonstrably greater than that of any of its constituent members. Here is not the place to enter a discussion of the mereological status of colonies or hives so it will have to remain undecided whether such things as ant colonies are unities or aggregates, mereological wholes or sums. We can leave the ontological status undecided yet still recognize the considerable potential niche value such things as insect colonies or hives have for the ecosystems in which they are situated: in short, it may not be necessary to make a judgment on the ontological status of the colony versus the individual. We might, however, need to treat a colony or hive *as if* it were a whole with a greater value than each of its constituent parts.

7. Feeling, Responsibility and Virtue

From the Nature of Things to an Ethical Standpoint

In this chapter we come back to the question of responsibility, but here the focus will be on recognition of responsibility and what happens when this responsibility is recognised by a moral agent.

As noted in the previous chapter, responsibility has at least a backward and a forward-looking aspect. Both have ethical implications, but it is the forward-looking aspect which affects most strongly the character of moral agents. Clearly, a virtue ethic can incorporate the concept of responsibility. If Rolston is right, however, pursuing a virtue ethic on its own, is 'dangerous'[1] and such an ethic needs to be complemented by a clear articulation that the source of value does not reside in the agent.[2] Responsibility in its forward-looking form can, at least in part, be seen in terms of virtue, but even if there is an 'imperative of responsibility' as Hans Jonas claims, acting on the imperative still requires there to be a good will which is 'not good because of what it effects or accomplishes' but is good without qualification.[3] Whether we take a virtue approach, an approach where the good will of an agent is necessary to act on an 'imperative', or some other approach, there still remains the question: how do we move from a statement about the nature of things in the

1 Holmes Rolston III, Environmental Virtues: Half the Truth but Dangerous as a Whole, Ronald Sandler and Philip Cafaro, eds., Environmental Virtue Ethics, Lanham MD: Rowman and Littlefield, 2005.

2 That a significant source of value in the world is outside the valuer has been demonstrated in the preceding chapters.

3 Immanuel Kant, Groundwork of the Metaphysics of Morals. Trans. Herbert Paton, New York: Harper, 1964, p. 61-62.

world to an ethical standpoint with respect to the things in the world? In their different approaches, Holmes Rolston, Emmanuel Lévinas, Martin Heidegger and Hans Jonas each offers a way to answer the question. Each is worthy of lengthy discussion that is not possible here, so the discussion below is merely a brief overview to indicate a range of possible answers. But before turning to them, it is appropriate to be reminded that the ethic of responsibility for nature presented here is based on a biological ontology that regards individual living organisms as self-realizing, self-maintaining entities with an internally-determined directiveness. Due to this internal determination each organism is a good-in-itself, but this, of itself, does not present sufficient reason for an agent to treat it as a morally considerable entity that makes a claim to be protected and have its interests promoted. So, how might we make the move from ontology to action?

One of the dangers of moving straight to a virtue approach is that in virtue ethics it is easy to place the origin of value within the valuer. That is, the primary locus of value would be in a (given) moral agent and only secondarily in moral patients or subjects. Rolston warns that we should not give a fundamental role to virtue in environmental ethics because, although acting virtuously toward the environment is a good state to be in, it is a mistake to think that human virtue is the source of value in nature [4] as living entities have value in themselves and, as Sandler summarises, 'environmental virtue is only intelligible as a responsiveness to the independent value of nature. After all, 'it is hard to gain much excellence of character from appreciating an otherwise worthless thing'.[5] For Rolston, focusing on human flourishing is a distraction that obscures the intrinsic value of natural entities as it is the intrinsic value, recognised by an agent, that makes environmental virtue possible. Rolston may be off the mark, however, in concentrating on virtue merely or primarily as 'an opportunity for personal development and flourishing' because although acting in

4 Holmes Rolston III, Environmental Virtues.
5 Ronald Sandler, Introduction, Ronald Sandler and Philip Cafaro, eds., Environmental Virtue Ethics, Lanham MD: Rowman and Littlefield, 2005, p.8.

certain ways toward the natural environment will tend to make an agent more virtuous, the attainment of virtue need not be the goal or intention of this behaviour. A virtue ethic need not 'endorse the claim that a virtuous person is motivated by a desire to become an exemplary individual'[6]. However, he is right that human virtue is not the primary locus of value. Character is developed in a dialectic of nature and culture and it would be a mistake to think that virtue in humans is the source of natural value as 'the value at stake precedes…our human presence.'[7] A human virtue may be generated, as the actualization of a human capacity, when a person respects an animal's life for what that life is in itself, but does this 'trigger' 'awareness of otherness' as Rolston claims[8] or is there something else at work here? Could otherness be an ethical category? Let us turn to Lévinas on this.

In turning to Lévinas we necessarily return to ontology. For Lévinas the Other has absolute primacy and if this is so, the way we understand the world—and our ethical relation to it—is turned on its head as 'being' would mean being 'for the other' rather than being for oneself. So, rather than a psychological 'ego' having ontological primacy (however one arrives at this) it is the Other that is always already there and it is only through the 'epiphany of the face' when we come face to face with a (concrete) other and 'recognise' it that we begin to understand who and what we are. Lévinas, using the resources of Husserl's phenomenology, attempts to explain 'the paradigmatic situation in which the Other as other 'comes to mind'' and in which 'I am bound to the others before being bound to my own body'.[9] The Other comes first, and from the moment the ego recognises an other he or she is 'summoned' to responsibility. But what if I choose not to respond to the summons? The summons can be

6 Simon James, Human Virtues and Natural Values, Environmental Ethics, 28, 2006, pp. 339-353.

7 Holmes Rolston III, Environmental Virtues, p. 72.

8 Ibid., p. 67.

9 Jean Greisch, Otherness: an ethical Category?, pp. 187-8

ignored, because after all I have a will, but the summons itself is part of my essence and to deny it would be to deny what I am —something that can be seen in Heidegger's later work on language.

There is in Heidegger's understanding of language an approach that Charles Taylor suggests could be the basis of a 'deep' ecology, a way of engaging with the natural world that is beyond a merely instrumental treatment and in which there is something beyond the human that 'makes demands on us, or calls us'.[10] For Heidegger we are beings in language: language is the house of being. Language not only 'speaks' (in itself), it also 'speaks' us. These are complex notions based in a power of words that standard theories of language can make no sense of and to which no brief exploration can do justice. However, in Heidegger's approach to language may be a possible resolution of the problem of how we might move from a claim about the world (e.g. that there is intrinsic value) to an ethical stance toward the world.

It is through language that 'world' is disclosed to humans. At the heart of this disclosedness is the concept of 'clearing' (*Lichtung*). The clearing is a space of expression that comes into existence only around *Dasein* (being-there). It is our Being that allows the clearing to happen, but it is not our doing, it is not within our individual power to bring it into existence, as it happens *between* speakers. And for it to happen between speakers, there must be 'speaking' itself as a precondition. But what might that mean for us here? It is known that Heidegger corresponded with Jakob von Uexküll and was well aware of the latter's work on the *Umwelt*. Von Uexküll, in fact, corresponded with Edmund Husserl as well as Heidegger and there is a demonstrable connection between the *Umwelt* of animals and the *Lebenswelt (Lifeworld)*[11] which Husserl defines as the historical, social and cultural world in which our daily life takes place. There

10 Charles Taylor, Heidegger, Language and Ecology, 1992, in Hubert Dreyfus and Harrison Hall, eds., Heidegger: A Critical Reader, Oxford: Blackwell, 2005, p. 246.

11 John Deely, Semiotics and Jakob von Uexküll's concept of Umwelt, Sign Systems Studies 32: 1/2, 2004.

may also be a connection—and here it may be sufficient to introduce this, but not expand on it at length—between what the linguist Thomas Sebeok refers to as von Uexküll's 'crypto-biosemiotics' and Heidegger's contention that (certain) entities demand that we acknowledge them. Charlcs Taylor interprets Heidegger to be saying that

> '...we can think of the demands of language also as a demand that entities put on us to disclose them in a certain way. This amounts in fact to saying that they demand that we acknowledge them as having certain meanings... this manner of disclosure can in crucial cases be quite incompatible with a stance of pure instrumentality toward them. Take wilderness, for instance. This demands to be disclosed as 'earth', as other to 'world'. [He continues to say that, from a Heideggerian perspective]... a purely technological stance' toward the natural world 'does violence to our essence as language beings.'[12]

Disclosure or revealing of world to humans takes place through language. But what of non-human living things? Taylor suggests that it was Heidegger's view that 'animals may have their own kind of clearing.'[13] In von Uexküll's notion of *Umwelt*, 'world' is co-created by a living thing's engagement with its external environment.

12 Taylor, Heidegger, Language and Ecology, p. 267.

13 Ibid., p.257n. Taylor says: 'This seems at least to be Heidegger's view [in Vorträge und Aufsätze]. Self-revealing and self-concealing in the animal are one in such a way that human speculation practically runs out of alternatives... Because the animal does not speak, self-revealing and self-concealing, together with their unity, possess a wholly different life-essence with animals'. It is worth noting also, the somewhat strange idea expressed by Plotinus that nature as such 'contemplates'. Suppose, he asks, 'that all things are striving after Contemplation... let us speak, first, of the earth and of the trees and vegetation in general, asking ourselves what is the nature of Contemplation in them, how we relate to any Contemplative activity the labour and productiveness of the earth, how Nature, held to be devoid of reasons and even of conscious representation, can either harbour contemplation or produce by means of the Contemplation which it does not possess.' Plotinus, Ennead III.8.3 in Plotinus, Psychic and Physical Treatises; comprising the second and third Enneads, transl. S. Mackenna, London: Philip Lee Warner, 1921, p. 120.

Although it is language not then used by von Uexküll, we can understand now that world is created through a semiotic engagement between an entity and that which is external to it. The speaking which is a precondition of *Dasein*'s engagement with and co-creation of world may be seen as a special case of the biosemiotic engagement through which living beings[14] create their *Umwelt.*

The clearing (*Lichtung*), the space of expression in which world is revealed to *Dasein*, arises 'between speakers'. But *Dasein* is also, at a fundamental level, an interpreter of signs —just as an animal is in its semiotic creation of its *Umwelt.* Are these really two completely distinct notions? If they are not distinct, but are related (it would seem one is a subset of the other), then *Dasein* will be called to pay attention to living things (in general) as things of value. That is, as the 'shepherd of Being' *Dasein* must *hearken*—listen attentively to and be instructed by[15]—the call made by living things and take care of (*pflegen*) them. Not to do so but to take an instrumental attitude toward the natural world would be to betray our essence as language beings.

Virtue is a disposition to respond well to demands from the world and not just to one's own interest. Following Heidegger, we recognise the demands of the world and respond to them by *hearkening.* But it is not just a matter of *what* we are, as it might be in a traditional Aristotelian approach, but it is also always a matter of *where* we are, of how we are situated *in* the world. Our essence is one of 'dwelling', which itself is a 'fundamental *relation* between humans and the world as 'being in' and 'being with'.[16] From the relationship to the demands

14 In von Uexkull's framework Umwelt is limited to bacteria, protista and animals. Fungi and plants are excluded.

15 Martin Heidegger, On The Essence Of Language: The Metaphysics of Language and the Essencing of the Word; Concerning Herder's Treatise On the Origin of Language, transl., Wanda Gregory Torres and Yvonne Unna, New York: SUNY Press, 2004, p. 94.

16 Christine Swanton, Heideggerian Environmental Virtue Ethics, Journal of Agricultural and Environmental Ethics, Springer: Published online, 17 June 2009.

of the world based on 'dwelling' it is possible to develop a virtues approach to ethics, as Christine Swanton does, but that is path that will not be followed here. Let it suffice that there is a 'call' from living things to which humans 'hearken' and which thereby presents them with a command to behave toward these living things in certain ways. If we listen to the call and act on it *well* we may be said to be behaving virtuously toward that which makes the call.

For Heidegger's student Hans Jonas the claim to be protected announced by a good-in-itself is met in humans by a responsibility that is an ontological fact, a fact that is within our being as *capacity* or *power*.[17] Simply because the power is embodied in humans means that each of us, as a member of the species, must attend to the being of other beings in such a way that we try to understand it, promote it and protect it. In being able to recognise a good-in-itself we face an imperative to look after it.[18] In effect, an entity that recognizes the good-in-itself of an other is placed under the (negative) obligation to prevent the entity being harmed and the (positive) obligation to ensure that it realizes its potential good. This is what amounts to behaving well toward it. The good-in-itself is, however, not a moral good. The moral good is recognized by a moral agent (where anything that has the power of autonomous action and the capacity to recognize the good-in-itself of an entity is a moral agent). In recognizing the good-in-itself of a self-realizing entity, the moral agent recognizes that the entity is a moral subject and establishes a relationship of non-reciprocal obligation between the entity and him or herself.

From this brief discussion it is apparent that there is in the living, non-human, world something that calls on us, which presents itself to us and demands of us that we engage with it in ways that value its value. How we engage with this something builds our character: we are beings who 'dwell' in a world where there is value in living things, in others of various sorts, and it would be a denial of our humanity,

17 Jonas, The Imperative of Responsibility.

18 Wendy Hamblet, To being or not to being, Anna-Teresa Tymieniecka, ed. Analecta Husserliana LXXXIV, 2005, pp. 357-364.

our essence, our being —in the many ways these can be described —if we were not to heed the call that is made.

Responsibility, Virtue, Feeling

So, moral agents are 'called' to behave well toward that which has value in and of itself. They listen or hearken to the call and act in various ways. If they act well toward that which calls they will contribute to their virtue. But *how* is the call recognised? In a Lévinasian world, when we 'face' and recognise a fellow human we see beings with value that announce a need —to which we (ought to) respond by asking: how can I help you? We understand *what* we are in the 'epiphany of the face'. An analogous understanding occurs when we encounter and engage with non-human living things. But at what level does this first occur? We engage semiotically with non-human living things in a manner much like reading body language in humans. In reading as 'signs' the various ways living entities engage with (and co-create) their *Umwelten* (of which we are part, because of the mutual semiotic engagement) we first *feel* something. There is then a response to the *feeling*. The response may be anything from revulsion to love in the first instance, but we are rational animals and subsequent responses are commonly enacted on the basis of rational judgment.

Responsibility as a Virtue that Responds to Feeling

However, let us look briefly at responsibility as a virtue before examining in more detail feeling itself as an originating impetus for moral action. Merely taking or accepting responsibility does not mean that a moral agent will always do the right thing in practical terms. For example, I have tried to rear orphaned magpies but they died because I fed them the wrong food; some of the trees I plant die despite my best efforts, although fewer die now than when I first started planting; and a friend rears orphaned kangaroos, but it took her years to learn how to do it well and in the early days some orphans died that today would survive. The point is that actions may be well-intentioned but wrong-headed. So what should be done?

We might start answering this by pursuing the idea that taking (substantive) responsibility is a virtue—a virtue that begins as a *feeling* in the moral agent and which, having arisen, is put into practice through the use of rational judgment. Aristotle was perhaps the first to have developed a theory of moral responsibility. In the *Nicomachean Ethics* he states that it is sometimes appropriate to lay blame or praise on a moral agent for what they have done or for some disposition[19] but only if the action or disposition is voluntary. That is, if the agent is aware of what she is doing and the action or disposition originates in the agent and is not compelled by some external force. Virtue in a moral agent, according to Aristotle, is cultivated. The capacity for virtue is present as potential—a power of the soul—but virtue needs to be practised in an intentional way so that over time it may become more nearly perfect. We have, by nature, the capacity for virtue and by putting the virtues into practice we learn how to be virtuous and build our character. 'The virtues,' says Aristotle, '...we acquire by first having actually practised them, just as we do the arts.'[20] It stands to reason that, in the same way that 'we become just by doing just acts, temperate by doing temperate acts, brave by doing brave acts',[21] we cannot be substantively responsible unless we take responsibility *for* something, that is, unless we *act* to protect and promote the interests of that thing. The Aristotelian virtues are

19 Aristotle, Nicomachean Ethics, transl. H. Rackham, Cambridge, Mass.: Harvard University Press, Loeb Classical Library, 1990, 1109b30-35.

20 Ibid., 1103a33.

21 Ibid., 1103b1-3.

dispositions,[22] that is, part of a formed state of character of a moral agent. One becomes virtuous through practice. In the case of responsibility, just as with other virtues, one learns how to behave in a manner consistent with being responsible, but *understanding* what it is that one should be responsible for is initiated, or triggered, by a feeling. The feeling may at first be *mere* feeling, i.e. it does not have a defined cognitive content, but we recognise that there is something *in* an entity (such as a baby) that requires us to consider how we ought to behave toward it. One must always already have the innate potential for responsibility (or care), and have that innate potential triggered into actuality, before one can cultivate it as virtue. A moral agent has a capacity or potential for responsibility that is actualized in response to the recognition that an entity has a claim of some sort on us. That is, the source of value is not human virtue, but something in an entity that we 'recognise' and respond to. A moral agent can always decide whether or not to accept or take responsibility, but if he or she does take on substantive responsibility—and necessarily try to behave in a manner that will promote the flourishing of the entity responded to — this is virtuous behaviour. Although none of the virtues is engendered in us by nature, we get from nature the *capacity* to include it in our moral make-up.[23]

Substantive responsibility in a moral agent is first triggered by a *feeling* the agent has for some object, a feeling that exists potentially in all moral agents as a necessary part of their make-up. It is a potentiality that is first actualized through practice and after being actualized is cultivated, developed, nurtured and grown through practice. A moral agent *feels* a responsibility to an object and then chooses (or not) to take responsibility for that object. Taking responsibility is clearly a virtue. How the moral agent recognizes a virtue and so becomes obligated in some fashion is an important

22 Ibid., 1106a13. Aristotle says virtues are a state of the soul and so must be an emotion, a capacity or a disposition. He rules out them being an emotion (1105b29) or a capacity (1106a6), so he says they must be a disposition.

23 Ibid., 1103a18-26.

distinction between the account of virtue proposed here and accounts of virtue in which a virtuous person is motivated by a desire to build an exemplary character.

Love of a Special Sort

The feeling that *triggers* virtue is 'love', but love of a special sort.[24] And it is this 'love' that allows a moral agent to recognize that he or she ought to take substantive responsibility for any teleological centre of activity, any entity with immanent purposiveness, in short, for any living individual. The sort of love referred to here is much the same as an aspect of love described by Spinoza, for whom 'love is a joy,' a property in the lover, 'accompanied by the idea of an external cause'.[25] Spinoza makes clear that love is not to be considered only a will of the lover, as that expresses a property of love and not its essence. He says,

> that when I say it [love] is a property in the lover, that he wills to join himself to the thing loved, I do not understand by will a consent, or a deliberation of the mind, *or* free decision...Rather, by will I understand a satisfaction in the lover on

24 Despite a similarity of terminology, this love is not the same as that which Harry Frankfurt talks of when he distinguishes passive and active love. For him 'love is…paradigmatically personal' and something the personal grip of which 'satisfies the conditions for autonomy that Kant believes can be satisfied only by the impersonal constraints of the moral law' (p. 132). People can appraise something as intrinsically valuable, however the 'things' that have intrinsic value for Frankfurt are not physical objects, but desires. Harry Frankfurt, Necessity, Volition and Love, Cambridge: Cambridge University Press, 1999.

25 Spinoza, Ethics, Trans. Edwin Curley, EIII Def. VI.

> account of the presence of the thing loved, by which the lover's joy is strengthened or at least encouraged.[26]

To draw this comparison with Spinoza's concept of 'love' is not to suggest that Spinoza himself would have wanted to allow that something non-human could be the 'external cause' of love. Spinoza was adamant that animals had no moral status and were owed no special consideration on their own account. Stephen R.L. Clark sees it as ironic that Spinoza's philosophy is employed by some recent environmentalist and Deep Ecology movements because, for Spinoza

> our highest good lies in contemplation of the whole and understanding our place in it… at the same time he [Spinoza] rules out any obligation to consider the needs or interests of non-human creatures.[27]

According to Clark, not only did Spinoza not feel it necessary to consider the interests of animals, he actually considered it wrong.

Spinoza's position on obligations to non-humans aside, the love or feeling by means of which a moral agent recognizes what she ought to care for, or take responsibility for, is not a consent or a deliberation of the mind or a free decision, but a 'joy accompanied by the idea of an external cause'. This joy or 'love' should be considered a state firmly rooted in the biological facticity of organisms—that is, this love exists because there are living things that are both the external cause of love and that of which love is a property. This 'love' should not be confused with the common senses in which we use the term: it is not erotic love (*eros*); it is not Christian love or charity (*agape*); and it is not brotherly love (*filias*). So what might it be?

Love in the sense used here is the act of opening up room beside us for the existence of others, allowing them to be, recognizing their claim to being. Love in this sense is a *potential* in all living things to recognize the being of others. This love is not, in the first place,

26 Ibid. , EIII Def. VI Exp.

27 Stephen R.L. Clark, Animals and Their Moral Standing, p. 88.

conscious: one becomes conscious of love when the conscious will determines to do something about a feeling. An organism does not need to be conscious to experience this love. Love as intended here is a biological phenomenon that allows recognition of others and permits coexistence. All animals, and perhaps all living things, have the capacity for this biological phenomenon of love. The feeling which in conscious moral agents gives rise to a sense of responsibility for others is made possible by the general biological phenomenon of love. Love in this sense is built into the very fabric of life and the ability of moral agents to act on the feeling made possible by this love is the actualization of a 'built-in' potential.

If one could 'recognise' something with an immanent *telos* in a manner analogous to the recognition of an other that happens when we are faced—come face to face with—a person, then one would enter into a moral relationship with it. There would be in the first instance a non-reciprocal relationship of obligation to consider the welfare of that something. If one were to enter into a virtue-based relationship one would become obligated to take (substantive, future-looking) responsibility for that thing's welfare. Of course it is possible to ignore the moral relationship, but this would be a betrayal of something essential.

Under an ethic based on the recognition of value in those things with an immanent *telos*, when moral agents are deciding what their moral duty is they are in the first place receptive to appeals to their emotions for them to act in certain ways. Once the appeal to their emotions has been made and recognized, moral agents may decide how to respond—but the feeling comes first.

Jonas' Ethic of Responsibility Revisited

One contemporary ethic that gives a central role to feeling and is predicated on moral subjects possessing an immanent *telos*—is Jonas' ethics of responsibility. When Jonas says that our emotional side must come into play and that our feeling is a necessary part of our moral nature, he is referring specifically to what he calls the feeling of (substantive) responsibility. An ethical theory dealing with responsibility should deal both with the rational ground of obligation and with the psychological basis from which an agent is moved to act. In Jonas' ethic of responsibility there are two sides: one rational or objective and the other emotional or subjective. It is the subjective or emotional side that is primary. This is so, he claims, because it is possible for emotion to be its own motivation to act, whereas

> '...no similar sufficiency can ever be enjoyed by the objective side: its imperative, evident as its truth may be, cannot become operative at all unless it meets with a sensitivity to the like of it...this sheer fact of feeling...is...the cardinal datum of moral life and, as such, implied in the 'ought' itself. It is indeed of the very meaning of the normative principle that its call is addressed to recipients so constituted that they are by nature receptive to it...the gap between abstract validation and concrete motivation must be bridged by the arc of sentiment, which alone can sway the will. The phenomenon of morality rests *a priori* on this correlation...feeling must supplement reason so that the objectively good can exert a force on our will...this insight lives in every doctrine of virtue.[28]

Despite the primacy that Kant's ethics gives to rationality there is a concession to the existence of feeling. However, while for Kant, emotion is important it plays a secondary role to rationality. One of the intentions of the present chapter is to turn this around so that rationality has an important place, but plays a secondary (but not necessarily subordinate) role to feeling. That is, *feeling*— and specifically the feeling of *love* as described above—is to be seen as an immanent motive force from which moral consideration is generated, whereas rationality (in the narrow definition usually assigned to it) is an *ex-post* tool for dealing with moral problems. In short, rationality is necessary for moral thinking, but it is neither primary nor sufficient.

28 Jonas, Imperative of Responsibility, p. 86

This runs counter to Kant's moral thinking in that while Kant does not deny that objects can directly affect us, he does deny that 'this emotive affection supplies the true motive for moral action'.[29] For Kant sentiment supplements reason. However, this sentiment is evoked not by an object, but by the sentiment of *reverence* for the moral law.[30] But imagine a scene where you come across an abandoned baby—whether human or other mammal. There is something in the 'object', the baby, itself that evokes an emotional response —an 'oh!' in the agent. At the point at which the agent recognises an 'other'—in Lévinas' terms comes face to face with it—there is a non-rational reaction. That (emotional) reaction is the first step to recognising the other as a 'thou' and so recognising a non-reciprocal obligation. But more of that later.

Jonas says only the object itself can evoke the feeling that initiates a moral relationship between agent and subject and it does this by evoking a feeling in the agent through its uniqueness as itself. What matters, in effect, is things in themselves, or the intrinsic value of things, and not the state of the will of a moral agent. It might be that, for moral agents, things can become ends only by engaging the will, but the will is initially spurred into action, so to speak, by the emotions.

It is possible to build on this and formulate an account of moral relationship which follows Jonas' to some degree and which stands on a notion of moral action directed to 'end-in-themselves.' Such an account would differ from Kant's in that, whereas for him a moral agent's actions are connected through reason to the recognition of an 'end-in-itself', in this new account there would be a direct connection through feeling. And while Jonas is correct that the object evokes a feeling in the agent through the object's unique 'self-ness', we should add a *caveat* that the moral agent must always already have the immanent potential to be affected by an object and to feel in certain

29 Jonas, Imperative of Responsibility, p. 86

30 In Respect for Nature Paul Taylor uses the notion of feeling in the same way so that emotion or feeling has a secondary importance.

ways as a result. For something to be a moral agent it must first have the potential to feel. That potential is actualised when confronted with an 'object'—such as a newborn—that announces a need to the world around it. So let us look more closely at what an account of moral relationship would look like if it fitted the brief sketched above.

The *feeling of responsibility* is a particular type of feeling that binds a moral agent to a particular entity (or moral subject) and prompts that moral agent to act to promote and protect the subject's interests. As such, *prima facie* at least, an ethic based on the feeling of responsibility appears to answer some of the criticism of Kant founded in feminist thought, in that it allows a primary role for feeling and involves a relationship between a particular moral agent and a particular moral subject—it is, in this respect, neither formal nor abstract. The feeling of responsibility presents itself as a strong basis for an ethic as it places agents under an obligation to act in the interests of those things recognized as moral subjects. The position here is precisely *not* that of Kant, who claims,

> However many natural causes, however many sensory stimuli there may be, which drives me to *will* something, they cannot produce [my state of] *being under obligation.*[31]

Contrary to Kant's position, the feeling of responsibility arising from the presence of natural 'love' does place an agent under an obligation to act. The feeling of responsibility is an objective principle based in a biological fact, according to which an agent ought to act.

It is an oft-repeated claim that modern ethical theory has been characterized by a rationalist bias that has precluded the recognition of the proper role of emotion or feeling in ethical matters. This absence of a role for feeling is one of the criticisms made by feminism and deep ecology alike, of modern theories of, particularly, environmental ethics. In her critique of Paul Taylor's, *Respect for Nature*, for example, Val Plumwood says Taylor aims to defend a biocentric (life-

31 Kant, Critique of Pure Reason, B576.

centred) ethical theory in which a person's true human self includes his or her biological nature, but that it attempts to embed this within a Kantian ethical framework that makes strong use of the reason/emotion dichotomy.[32] She goes on to suggest that Taylor treats 'care', viewed as inclination or desire, as irrelevant to morality and that Taylor's rejection of a primary role for feeling is reflected in modern ethics generally because of an assumed superiority of rationality. Plumwood wants a more central role for the concept of *care,* and for the virtues generally, and suggests that the virtue of *care* is one that should be nurtured. It should be noted, though, that the virtue of care does not describe the full complexity of care. There are at least three elements to care: care for, care about and taking care. These are complex concepts which can only be touched on here. Caring *for*, as a well-intentioned act, may enable an agent to demonstrate the virtue of care. Caring *for* acts transitively: there must always be an object of caring and the carer must be engaged in the act of caring. And caring *about* something or someone may be a necessary condition to be able to demonstrate the virtue of care, but it is not sufficient since caring action is required. The matter of caring *about* something is taken up by Harry Frankfurt in a number of places. For instance, in *Necessity, Volition and Love* he notes that

> To care about something differs not only from wanting and from preferring it but also from judging it to be valuable. A person who acknowledges that something has considerable intrinsic value does not thereby commit himself to caring about it. Perhaps he commits himself to recognizing that it *qualifies* to be *desired for its own sake* and to be *pursued as a final end.* But this is far from meaning that he does actually desire or seek it, or that he ought to do either.[33]

Taking care (in the sense of paying attention to detail or in the Buddhist sense of mindfulness) may likewise be necessary, but not sufficient for the virtue of care. Taking care—or being careful, in

32 Plumwood, Feminism and the Mastery of Nature, p. 166.
33 Harry Frankfurt, Necessity, Volition and Love, p. 158. As noted earlier, Frankfurt is concerned with the value of desires.

itself, may be more prudential than virtuous. All three elements of care may be needed to exercise the virtue of care, but there is still something missing.

That something is revealed when we ask ourselves whether we are giving merely what we want to give or whether we are seeking an answer to the question: what can I do to help? An example of the first might be dropping a few coins at the feet of a down-and-out in an automatic gesture, whereas in the second case we 'face' a moral subject, engage with the subject and seek to find out what that subject needs. In so doing we acknowledge an as-yet undefined obligation and a sense of substantive responsibility. But to whom or to what might we have a substantive responsibility? Is there a case, as John Llewellyn argues, for 'not limiting to human beings the range of beings to whom or to which underivative responsibility is due'?[34]

There is much in common between *care* and *responsibility*. The newborn, for example, imposes an obligation of 'ought-to-*care*' on its parents and also on any other moral agent that can help it. This is an obligation that an agent has to a *particular* other and not to others in general. This obligation is expressed similarly when we recognise that something facing us is a 'thou' —a singular other with the status of a moral patient —and we fall under an obligation to ask, in effect, 'How can I help you?' The obligation is non-reciprocal. But how might this obligation come about? For Lévinas, the 'Other' is not knowable in itself but reveals itself to us in the face-to-face encounter with an other—a concrete other to which we can point, an other which 'faces' us. The revelation of the Other (through encounter with a 'face') makes a demand on us. This demand is more than a mere asking, it imposes an obligation. But it also has the effect of founding our being-in-the-world because it gives it meaningful direction: attention to and action in aid of an other. Responsibility for the other is always already there—before we realise it. We always already have an obligation to

34 John Llewellyn, Emmanuel Lévinas: the genealogy of ethics, London: Routledge, 1995, p. xi.

anything that has the status of a thou (second person singular other). On recognising a 'thou' an *a priori* obligation to Other is triggered and we are obliged to ask how we can help. There is 'a responsibility for the Other ... before commitments have been entered into, a responsibility that is the *a priori* measure of all commitments, rather than being *a posteriori* measured by them'[35] For Lévinas human identity happens through responsibility to the other. However, despite Lévinas' thoroughgoing anthropocentrism and the likelihood that he would not necessarily like the idea of appropriating his thought in this way, we should still ask whether only humans can be singular others for whom we take on responsibility in facing them. If we were to take Llewellyn's position with regard to animals,[36] for example, then the singular other —or individual moral patient —that places a demand on us simply because it exists, need not be limited to humans. But let us leave that open —as an invitation for it to be possible. [37]

Responsibility takes multiple forms. For example, it can be viewed as responsibility 'taken' or responsibility 'assigned'. It is taken in the sense that a moral agent freely chooses to take on responsibility for something; it may assigned if the responsibility is given to a moral agent by someone or something. Legal responsibility would fit in the latter category. It might also be viewed in Jonas' terms as either formal or substantive —although this distinction might be better viewed in terms of 'past' (I did 'x' to 'y') or 'future' responsibility (I will ensure that 'x' will —or will not —happen to 'y'). Care in its turn might be seen as having two major forms, which could be called the active and the passive. The passive sense of care operates when an agent has some interest in the welfare of an object, or the success of

35 Zygmunt Bauman, Postmodern Ethics, p. 85.

36 John Llewellyn, The Middle Voice of Ecological Conscience, London: Macmillan, 1991.

37 We might also speculate that Lévinas' position may be the basis of Jonas' claim that the 'ought-to-be' of an object comes first in an ontological sense and that this is necessarily followed by an 'ought-to-do' of the 'subject who, in virtue of his power, is called to its care.'

an enterprise, but when it is not necessarily in the agent's power to affect the outcome. In this sense of caring, I might care whether my favourite football team wins on Saturday, or care whether there is a lasting peace in the Middle East. This can also be called 'caring *about*'. Just as the substantive sense of responsibility operates when an agent chooses to take responsibility *for* a particular object and has the power to do so, the active sense of care operates when an agent looks after, nurses, nurtures, or in other words cares *for,* an object. This is not to suggest that either care or responsibility collapses into the other—or even that they should be considered much the same. There are similarities between care and responsibility, but differences also. For example, caring *for* generally involves more of a face-to-face relation than responsibility and generally requires a more hands-on approach. Care is a much more complex concept than the preceding brief treatment can address and, although justice has not been done to it here, it does have an important role to play. And emotion is clearly more than a mere prompt to start a rational process.

From the above it should be clear that one way to overcome a rationalist bias in ethics is through a virtue ethics in which something like a virtue of responsibility arises from feeling rather than being contingent on some form of rationality, narrowly conceived. In short, a primary role should be given to virtue grounded in feeling and only a secondary role to reason. Plumwood too wants an approach to ecological morality that moves 'in the direction of virtue' as this would avoid 'ethics in its usual rationalist conception'. She suggests an ethics of virtue should be based in concepts such as 'care, respect, gratitude, sensitivity, reverence and friendship' as these concepts are more resistant to analyses that make an arbitrary division between reason and emotion; they are more local, allowing for a focus on particulars, and they do not need reciprocity. The concepts are also moral 'feelings' that involve 'cognitive elements, ethical elements and emotion in ways that do not seem separable.' The virtues create a relationship of one-way obligation: the moral agent feels obligated to assist a moral patient, but there is no corresponding obligation for the moral patient to act in any way toward the moral agent. Responsibility likewise is a non-reciprocal relationship between a particular moral agent and a particular object of responsibility; it is not an abstract concept and would satisfy the feminist principle that any acceptable morality be 'grounded in sound particularistic relations to others in personal life'.[38]

Shortcomings in Jonas' Ethic of Responsibility

Jonas' ethic of responsibility seems at first glance to answer some of the criticisms made of contemporary ethical theories, but it too has problems. In fact, there are a number of serious shortcomings with his position. Before examining the problems, however, a quick review of Jonas' ethic is called for. Firstly, Jonas adopts an Aristotelian ontology—a biological ontology in which organisms have an immanent *telos* and a good of their own. And, secondly, he takes a biological rather than an ecological approach to ethics. As demonstrated earlier, in Chapter 3, this is an approach that is necessary, not least because of the difficulties in establishing the ontological status of ecological systems i.e. whether they are wholes, individuals or aggregates. Jonas recognizes individual organisms as the basic unit of moral considerability and recognizes that they strive to persist as what they are. That is, in the terms discussed in earlier chapters, they are self-realizing, self-maintaining entities and as such they have immanent purposiveness, a self-reflexive active *dūnamis,* a *conatus.* In Jonas' ethic the principle of responsibility entails a *positive* duty on moral agents in that there is a moral *ought* generated that requires a moral agent to *act* to protect and promote the interests of a particular moral subject. It is not enough that a moral agent should simply *consider* a moral subject, the agent also must *act* in the subject's interests.

38 Plumwood, Feminism and the Mastery of Nature, p. 183.

There are two main points in Jonas' ethic of responsibility that should be reconsidered. Firstly, although Jonas challenges the ontological assumption in modern philosophy that there is a sharp divide between human beings and the rest of nature, he puts humans at the top of a hierarchy of moral considerability and gives moral priority to humans without acknowledging that there might also be another value, such as an environmental value, that should be taken into account. As Jonas portrays it, humans take (substantive, future-looking) responsibility primarily for humans, and there is responsibility for other living things only to the degree that they are necessary to the continued survival of humans. As much as it may be psychologically and politically important, such human chauvinism is, however, not a necessary part of an ethic based in responsibility as a concept of moral responsibility can effectively cover responsibility for nature and responsibility for humans, precisely because humans are part of and not apart from nature. Secondly, although Jonas' imperative of responsibility recognizes that there is value-for-its-own-sake in all living things and that this can provide a base on which to build an ethic, there is built in to his position a totalitarian politics that actually amounts to an abrogation, or at least a denial, of responsibility —for all except the executor of supreme power. Those who are subject to the power wielded by a dictator—even the dictator of a green fascism—lack the freedom to choose what course of action they should take. And, because they lack the freedom to choose, they may be unable to exercise the *virtue* of responsibility, it being replaced by something like a *legal duty* of responsibility for nature.

Because he insists on priority for humans and says that humans should look after non-human nature in order that humans may prosper, Jonas' theory becomes in effect a secular stewardship having much in common with the Christian stewardship model of environmental ethics[39] but without the divine authority characteristic of that model.

39 The locus classicus of the stewardship model of environmental ethics is John Passmore, Man's Responsibility for Nature. For a survey article which includes discussion of stewardship, see e.g. J. Baird Callicott, The Search for an

This anthropocentric bias, which amounts to a human chauvinism, is not a logically necessary element of an ethic of responsibility. In Jonas' case, the anthropocentric bias appears to be merely an *ad hoc*, additional assumption.

Jonas argues for replacing the imperatives of what he calls the old ethics, such as Kant's categorical imperative, with the prime imperative of his new ethics—an imperative that can be expressed in the maxim: 'act so that the effects of your action are compatible with the permanence of genuine human life'.[40] However, Jonas' imperative of responsibility with respect to organic nature actually amounts to an argument which runs something like this: man must survive; if man is to survive then nature must survive and if nature is to survive then man must have an attitude of responsibility toward it. His maxim is also at odds with his key claim that moral agents have a responsibility to the perishable *qua* perishable. If prime responsibility is owed to that which is perishable then the objects of that responsibility would not be limited to humans, but would extend to all living things—as all living things, and not just humans, are perishable.

One example he gives of this responsibility to the perishable is in relation to the biosphere, 'an object of an entirely new order...added to what we (humans) must be responsible for because of our power to act over it'. The biosphere is to be considered perishable and hence an object of responsibility. But Jonas' position begs the question of whether the biosphere has value in itself or is the sum of values of its constituents plus some value (potential or actual) in the interrelation of the constituents. If the biosphere itself is a morally considerable entity,

---

Environmental Ethic, Tom Regan ed. Matters of Life and Death. Christian stewardship basically sees it as the God-given responsibility of humans to look after the earth and all its creatures, often stemming from readings of the Old Testament such as at Genesis 1:26 where God gave man 'dominion over the fish of the sea and over the birds of the air and over every living thing that moves upon the Earth' and at Gen. 2:15 where God 'took the man and put him in the garden of Eden to till it and keep it' (The Holy Bible, Revised Standard Version, 2nd ed. London: Oxford University Press, 1971).

40 Jonas, Responsibility, p. 10.

and human responsibility for the biosphere comes from human power to affect it, why, in principle, should humans' prime responsibility be to humans and not to the biosphere? And if humans cannot affect the biosphere without first affecting individual living components of the biosphere, as argued in Chapter 3, why should human responsibility not be primarily to the individual living things over which humans have influence?[41] Jonas thinks he has an answer to these questions. He claims that the whole of nature has become a human responsibility and calls for a new ethics that extends 'the recognition of 'ends in themselves' beyond the sphere of man and makes the human good include the care of them'.[42] To this it might be added that human responsibility for the whole of nature is both formal and substantive, and both backward looking and forward looking, in that humans have contributed to the degraded state of much of the natural world and also have obligations to remedy what can be fixed and to protect from damage entities with a good of their own. In Jonas' account, however, the care of non-human 'ends in themselves', or living things, should be undertaken for the benefit of humans and not for the benefit of the living things themselves. This human chauvinism has its roots in Jonas' claim that he has found the answer to 'the critical point of moral theory, namely *what passage leads from willing to*

41 In terms of environmental pragmatics, however, it may not matter whether an ecosystem is a unity, an individual in ontological terms. This is because, even though an ecosystem might have no standing as an individual, ontologically speaking, it might be made a 'thou' (a second-person singular entity) for political purposes and so be the object of moral consideration and responsibility. It would, in effect, become an honorary 'thou', an honorary individual moral subject. Identifying an ecosystem or even 'the environment' in this way would simplify policy making as decisions affecting the welfare of individuals within ecosystems, such as wilderness areas, could be made as if the ecosystem itself were morally considerable. This has an immediate parallel in the legal fiction of a person. An ecosystem would, for policy purposes be similar to a legal person whose interests there is an obligation to protect. The responsibility in this case might be a legal responsibility, but wilderness would not know the difference.

42 Jonas, Responsibility, p. 8.

obligation'.[43] The transition from willing to obligation, he says, is 'mediated by the phenomenon of *power* in its uniquely human sense, in which causal force joins with knowledge and freedom.' He acknowledges that 'power' as purposive causal strength is in evidence throughout much of animal life, but claims that in non-human nature this power is 'blind' even though it is driven by purpose. This blind 'purposive causal strength' amounts to the same thing as that referred to in previous chapters as an active *dūnamis* and is consistent with the immanent purposiveness that in an Aristotelian biology is a defining property of a living thing. There is a strong link between will, responsibility and *dūnamis* in that it is having the choice of whether to exercise one's power (active *dūnamis*) or not that allows for a moral agent to take substantive responsibility for the wellbeing, or flourishing, of a moral subject. This effectively limits Jonas' second form of responsibility (substantive, forward-looking) to those entities that have an active, conscious *dūnamis* i.e. those that bring together power and knowledge. But, this does not necessarily limit responsibility to humans, contrary to Jonas' claims. For example, the higher primates (at least) exhibit intentional behaviours that are other-regarding.

Not all the purposive causal strength in non-human nature is blind. And, even though the power of any one living thing is restrained or contained by the counterplay of other living things also carrying on their natural purpose, this does not mean, as Jonas maintains, that it is only in humans that power is emancipated from the whole through knowledge and arbitrary will. It is not the case that only in humans does there arise 'out of the willing itself the 'ought' as the self-control of his consciously exercised power'.[44] Even if this were the case, and as a result humans should be considered the only moral creatures, it does not explain why humans ought to be the prime *objects* of responsibility.

43 Ibid., p. 129. Jonas is apparently addressing here Kant's argument at Critique of Pure Reason B576.

44 Ibid., pp. 129-130.

To explain why the prime responsibility should be to and for humans, Jonas relies on the idea that 'man' is the highest peak of the principle of purposiveness. That is, all living things have the property of purposiveness, but it is in humans that the property is at its most complex and can exert the greatest influence. It is because 'man' is the highest peak of purposiveness, a peak reached through the freedom to set ends and the power to carry them out, that Jonas says man should be the first object of his own obligation. After the prime responsibility has been met subsequent responsibility comes from being 'custodian of every other end-in-itself that ever falls under the rule of his power'.[45] This amounts to an (insufficient) justification for a self-interested environmental stewardship. As argued earlier, complexity is important, but so is niche value—ecological, semiotic and ontological. Humans may have a claim to priority based on their greater complexity and greater capacity for semiotic engagement with their world, but this might be counterbalanced by other values.

One of the more serious drawbacks to Jonas' version of an ethic of responsibility is that implicit throughout is a preference for a totalitarian politics as the way to ensure that the interests of the biosphere are protected. He says, for example, that the 'mere fact of superior power legitimates its use' and, with reference to environmental crisis, that 'only a maximum of politically-imposed social discipline can ensure the subordination of present advantages to the long-term exigencies of the future'.[46] His preference for a totalitarian politics is made explicit in Chapter 5 of *The Imperative of Responsibility* where he says that a socialist society, especially one that centralizes power and concentrates decision-making at the top, is best suited to deal with the 'threatening future'. He says

> What we are talking of so far are the governmental advantages of any tyranny, which in our context one must hope to be a well-intentioned, well-informed

45 Ibid., p. 130.
46 Ibid., p. 137.

tyranny possessed of the right insights...[and] only an elite can assume, ethically and intellectually, the kind of responsibility for the future we have postulated...[47]

In other words a form of totalitarianism is to be supported provided it is 'green' or environmentally aware.

However, quite apart from the independent arguments one might mount for or against such a position, a totalitarian politics is not entailed in the adoption of an ethic of responsibility. Personal, individual responsibility by moral agents for moral subjects is, in fact inconsistent with a totalitarian politics because even the best-intentioned totalitarian system at some level denies moral agents the ability to choose for themselves what actions to take. In such a system moral agents would—in some situations at least—be coerced into complying with instructions dictated by the ruling elite and, since moral decisions can be made only when a moral agent is free to choose what he or she will do, this could lead to moral agents being denied the ability to make moral decisions. Responsibility is, in its primary form, a private relationship between an *individual* agent and an *individual* or particular 'other'. The political responsibility that Jonas envisages involves an individual or group of individuals taking power in order to take responsibility for others. The intended primary consequence, of protecting something, may be worthwhile, but an authoritarian approach is not the only way to ensure that the perishable *qua* perishable are protected and it should not be the first choice of methods as so much of value is lost in the process. If Jonas' arguments that political responsibility should ultimately rest with a 'statesman' who takes on ultimate power so as to exercise ultimate responsibility were to be accepted into an environmental ethics, then the result, at best, would be a form of green fascism in which 'the mere fact of superior power legitimates its use'.[48] This violates the generally accepted principle that all humans are moral equals, entitled to equal consideration as both a moral subject and as a moral agent.

47 Ibid., pp. 146-147.
48 Ibid., p. 137.

In contradistinction to Jonas' position, however, political responsibility can be seen as natural responsibility made public and generalized. All moral agents are capable of exercising natural responsibility—and under a non-totalitarian political responsibility private thinking would be capable of being made public and so it would be possible for a consensus to be reached among groups of like-minded people. This view of political responsibility would allow for a politics based in local action, for small groups of individuals to come together, to make consensual decisions and then, through either collective or individual action, for each to make a (positive) difference to the individuals in their immediate environment by taking (substantive) responsibility for their well-being or flourishing. Each individual would be free to make moral decisions and free to act in concert with like-minded fellow citizens. In short, an ethic of responsibility predicated on the moral value of beings possessing an immanent *telos* can be translated into non-totalitarian, localized, political action. Such a system would be local, would value what is perishable, would be able to employ methods appropriate to local conditions to achieve desired results. It happens, for example, in the village where I live. Small local groups get together to look after reserves of natural bushland, or work together to clear streams of exotic plant pests so that indigenous species (plants and animals) can re-establish themselves. Some of this work is organized under the umbrella of the local authority, but much is simply organized and co-ordinated by locals getting together, agreeing on what needs doing, and setting out to get things done. What is necessary for this to happen is good will, a sense of responsibility and the determination to act—the very qualities governments should exhibit. Of course, such local good will is not always present and good will may be overridden by economic imperatives to eat, drink and be housed, but noting its lack does not mean it ought not be promoted through education and policy.

Whether as a personal virtue or as public virtue, the foundation of responsibility for nature is moral agents recognising the demands made on them by (living) things that have value for themselves. The demand is an imperative: for if we deny the demand we deny something essential about ourselves. It may not matter whether we address the demand as individuals, locally in concert with others or through large-scale political action, as each of these is needed. Moreover, not to act, not to value through our actions that which has value in itself is the opposite of virtue.

8. Finally

There is a common thread running through some of the main biocentric theories of environmental ethics, a thread that is more than just the fact that they are life-centred. This common thread has two main strands: the possession of an immanent *telos* as a necessary property of living things and the claim that entities possessing this immanent *telos* both deserve moral consideration and, in effect, demand it of those moral agents whose causal powers can affect them.

This book demonstrates that immanent, or in-dwelling, purposiveness is a defining property of all and only living individuals—and is a purposiveness that needs to be protected and allowed to flourish.

In an ethic predicated on an Aristotelian biological ontology, an organism is the paradigm case of a morally considerable individual. Morally considerable individuals can variously be said to be 'selves', to strive to persist in their being, or to possess immanent purposiveness. On the question of selfhood, it is clear that individual organisms are highly organized, metabolizing aggregates that set themselves off from and actively trade with the environment—in effect, they co-create their environment, or *Umwelt*. The attainment of the state in which organisms are simultaneously self-dependent and other-dependent, constitutes the horizon of selfhood.[1] Every living thing is a mode of being and striving, and every feeling and striving being is not only an end of nature but is also an end-in-itself, its own end. Immanent purposiveness is an integral and necessary part of nature, life itself is *a* purpose of nature and 'purposive being' is the basic purpose that life brings to fruition.

1 Jonas, Responsibility, p.73-74.

Individual organisms are morally considerable entities because they have an immanent good, a good-in-itself (and so have a *conatus* or *dūnamis*—an immanent motive principle). One consequence of this is that value in nature is to be taken as the sum of the values of individuals. Ecosystems, on this account, are held not to have value *in themselves* over and above the sum of the values of the individuals in the ecosystem, but individual organisms have an ecosystemic value in addition to whatever value they have in themselves. The value of something will vary with the actual interconnections between it and other elements of the ecosystem or community. That is, ecosystemic value supervenes on the value of the biological individuals that inhabit the ecosystem, so that organisms have, in effect, a core value coupled with an environmental, semiotic and ontological niche value. An organism's total value at any one time is the sum of its core (or essential, or intrinsic) value plus the value that it has to others in its environment.

Adopting an ethic built on responsibility for those individuals with a good of their own does not prevent us from taking a thoroughly pragmatic stance on how actually to take substantive responsibility. It does not, for example, stop us from assigning to ecosystems a *political* value that can be employed to achieve political ends. It is just that to ascribe the same ontological status to ecosystems as are ascribed to individuals and on that basis to ascribe the same sort of value is to make a category error.

In an ethic of responsibility, it is the *prima facie* duty of moral agents to preserve the immanent, self-realizing good within those entities that have inherent worth. In other words, under an ethic of responsibility, a living thing ought to be looked after or cared for simply because it exists as a living thing. A living thing can never be seen as a 'mere' *is*, but always addresses an *ought* to those around it. The 'ought' is addressed whether it is recognized or not. At the point the 'ought' is recognized, a relationship of non-reciprocal obligation is established between the addresser and the addressee, between moral subject and moral agent. But who can recognize the 'ought' and how is it that they can come to recognize it? The 'ought' is recognized, in the first instance as a *feeling*, by moral agents who come into contact with the living thing. A living thing is a moral subject and as a moral subject it presents a demand to any moral agent that recognizes it as a moral subject. This demand requires that the moral agent take substantive responsibility for the moral subject and ensure that the moral subject can realize its potential good, its good in itself, for its own sake. That is, the move from an 'is' to an 'ought' happens via a feeling in the moral agent, a feeling that may be characterized as *sympathy,* where sympathy is understood as Stephen Darwall describes it, as:

> a feeling or emotion that (a) responds to some apparent threat or obstacle to an individual's good or well-being, (b) has that individual himself as object, and (c) involves concern for him, and thus for his well-being, for his sake.[2]

Sympathy in Darwall's view is felt from the perspective of 'one-caring: the key issue is that concern is *for the other's sake*. This concern or desire for something 'for someone's sake' is a 'a form of desire that springs from care' whereas imaginatively sharing a person's concerns (such as one would experience in empathy) 'is hardly the same as sympathy *for* her'.[3]

If we expressed the foregoing in terms of *rights,* moral subjects would have both a liberty right and a welfare right. That is, they all have the liberty right, the right to be left alone, simply to pursue their own flourishing or well-being, to pursue their own *telos*, but they also have a right to assistance from any moral agent capable of giving them what they need to pursue their *telos.* There is an onus on moral agents not only to *allow* moral subjects to be what they are and to actualize their potential but also actively to *assist* them to do so when this is needed. Viewed in this way, one might exercise responsibility for all the individuals that make up a forest, for example, simply by leaving

2 Stephen Darwall, Empathy, Sympathy, Care, Philosophical Studies, 89, p. 262.
3 Ibid., p. 263.

them alone. There would be no need to intervene on their behalf unless to counter some external threat.

It should be clear that the *prima facie* duty to consider moral subjects suggested by both Taylor and Mathews can be fulfilled by taking responsibility for those subjects, where taking responsibility is a stronger (more active) approach than that of merely showing respect or consideration. Responsibility is a response to an obligation a moral agent has simply because the moral subject to which responsibility is to be directed exists. That is, where a free will meets a living being (which has its end in itself), *is* implies *ought*. The free will ought to take responsibility for the living thing and act in its interests because, simply by existing, the living thing makes a claim on moral agents to assist in the realization of its immanent good—or, if not actively to assist then at least to ensure that it is not prevented from realizing its good. And for a moral agent not to act on the claim or demand made would be to deny part of his or her essence.

Humans ought to have responsibility for nature, but not a responsibility narrowly conceived and not a responsibility simply to ensure wise stewardship so that the bounty of nature can be enjoyed by humans. Enjoying the bounty of nature might well be a desirable secondary outcome, but it should not be considered the prime moral motivation. That moral agents ought to take responsibility for all the individual moral subjects in nature is implicit in key contemporary works on environmental philosophy. The aim in this book has been to make this responsibility explicit without falling into a stewardship model or a model that leans toward green fascism.

Although care should not be equated with responsibility, responsibility for nature is linked closely to an ethic of care for nature as the two concepts are closely allied and share considerable common ground. One significant similarity rests on their mutual relationships to moral imperatives. When one recognizes that one has a moral duty of care toward beings with a good of their own—as Mathews does—, the voluntary attitude of caring is transformed into a moral imperative, just as Jonas' second form of responsibility becomes an imperative because of the 'ought' implicit in the demand for intrinsic good to be realized.[4]

But how do we begin to recognise the imperative? For Taylor, 'our character expresses respect for nature when it enables us to see clearly what those duties, obligations, and responsibilities are and to carry out their requirements'.[5] Taylor requires an agent to act, but the act can only be moral if the end of the action is ethically obligatory, i.e. if it conforms to a rule or set of rules that embody the attitude of respect. For Taylor, moral behaviour is necessarily and primarily rational. This book, on the other hand, argues that an emotion or a *feeling* comes first. For Taylor, the *feeling* of reverence is for the moral law—as it was for Kant—whereas in an ethic of responsibility for nature moral agents recognize the object of feeling to be a particular living organism itself and not an abstraction such as a moral law.

In contradistinction to Taylor, there is no need to distinguish between a human ethic and a nature ethic: there need be just one ethic for all living things because all living things are *prima facie* morally considerable. This is not to say that all living things are equally morally considerable. However, although more complex organisms are, other things being equal, entitled to more favourable treatment than less complex organisms, the final moral equation must take into account the value the organism has to the ecosystems in which it persists. That is, the final moral judgment must give due recognition to the *niche value* that the organism—as a representative of a

4 We should also note that for Frankfurt, (Necessity, Volition and Love, p. 133) care is a form of love, particularly active love, and where 'love need not be based upon self-interest. It may be fundamentally active, differing from passive love in the nature of the lover's motivation and in his concern for whatever it is that he loves…In active love, the lover values this activity for its own sake instead of for the sake of advantages that he himself may ultimately derive from it…his primary goal is not to receive benefits but to provide them. He is motivated by an interest in serving the interests and ends of his beloved rather than by an interest in serving his own'.

5 E.g. Taylor, Respect for Nature, p. 88.

species—has, where niche value is inclusive of an ontological and a bio-semiotic value. The ethic of responsibility is an ethic in which moral agents have an obligation to take responsibility for all moral subjects (i.e. all living things) which they will actually or potentially affect through the exercise of their causal powers. All living things stake a claim to moral consideration simply because they exist and whenever their interests are likely to be affected by the actions of a moral agent, the moral agent has an obligation to take substantive responsibility for the protection and nurturing of those interests. This is not to say, however, that there are no competing obligations or that a calculus of some sort may not be needed in working out pragmatic solutions to conflicting or competing obligations.

Under an ethic of responsibility, the most complex organisms have the strongest *prima facie* claim to moral consideration, but when there are conflicting demands for moral consideration the value a thing has to its environment must be taken into account. In effect this means that those organisms that contribute most to the permanence and stability of the ecosystem have a greater claim to be considered—that is, they impose a greater claim to be the objects of our responsibility—because the permanence and stability of ecosystems is necessary for the maintenance of the various entities that seek to achieve or maintain a good of their own. In short, this amounts to an ethic of responsibility coupled with something like Leopold's land ethic, in which

> a thing is right when it tends to preserve the integrity, stability, and beauty of the biotic community. It is wrong when it tends otherwise.[6]

Of course, it is not always immediately clear whether an action will preserve the integrity, stability, and beauty of the biotic community or whether one action will be more successful than another in achieving this aim. A set of principles is needed to supplement the land ethic

6 Aldo Leopold, A Sand County Almanac and Sketches from Here and There, New York: Oxford University Press, 1987, pp. 224-225.

principles so that in practice the best result is achieved. These principles will serve as a guide so that responsible moral agents help realize the greatest good possible.

The imperative of responsibility can be coupled with a consideration of the ecological 'fit' (i.e. niche value or degree of interconnectedness) of individual organisms to arrive at a set of workable principles to guide practical judgment. Taylor's rules of duty (two negative duties: the rule of non-maleficence and the rule of noninterference and two positive duties, the rule of fidelity or nondeception and the rule of restitutive justice—which 'imposes the duty to restore the balance of justice between a moral agent and a moral subject when the subject has been wronged by the agent'[7]) can readily be replaced by the principle of responsibility since, if one takes on responsibility for a moral subject, one will not act to harm or needlessly interfere with that subject, nor will one normally be unfaithful or deceitful. One would not *normally* be unfaithful or deceitful because it is conceivable that one might deceive in order to protect an organism's vital interests. For example, to save an endangered marsupial one might need to transfer it to another habitat and to transfer it one needs to trap it. And, of course, trapping is a form of deception.

When it comes to resolving disputes, something other than Taylor's five principles (self-defence, proportionality, minimum wrong, distributive and restitutive justice) is needed. Taking the niche value of organisms into account, for example, provides a basis for determining the priority to be given to particular organisms at a given time in a given place. That is, when making decisions that affect ecosystems or communities, taking into account the niche value of an individual (or what Mathews would call their degree of interconnectedness) makes a just decision more likely because the protection of high niche-value organisms is likely to lead to long-term benefit for an ecosystem. However, the inclusion of niche values

7 Taylor, Respect for Nature, p. 186.

involves a complex calculus that needs to take into account the value of an organism to all of the other organisms it affects or has the potential to affect. That a completely accurate calculus is unlikely to be practicable should not prevent the calculus being attempted. The need for an accurate calculus indicates very clearly that sound data from well-conceived ecological science is one of the necessary conditions for good decision-making.

It is worth also asking whether there needs to be special consideration given to the moral value of moral agents, as this would need to be included (somehow) into decision-making. It can be argued, for example, that, since only moral agents can take responsibility and since taking responsibility is desirable, not least for the benefit of others, that moral agents are in a special category requiring special moral consideration. That is, moral agents require special consideration because moral agency itself should be preserved. But, if there were no moral agency in the world, would the world be a better or worse place? If there were no moral agency, there would still be the biological phenomenon of *love* underpinning all actions by all organisms. Harmony would prevail, even if it were simply the dynamic equilibrium that all complex systems must have merely to persist over time. However, one of the characteristics of moral agency is that it has the capacity to be a force for good. Moral agents can add to the stock of good in the world—even if a categorically different type of good to that inherent in all living things simply because they are living. Because moral agents can add to the stock of good, the earth should be a better place with them than without them—even if they make well-intentioned decisions that unfortunately affect adversely the interests of some moral subjects. On balance, it would have to be a better place with moral agency than without and moral agents consequently have a greater claim for moral consideration than do mere moral subjects.

Aristotelian biological ontology provides a basis for a biocentric ethics. The immanent *telos* that all living things possess, this end-in-itself, 'claims' value to and for itself and so value enters the world, ontologically speaking. This basic value that all living things possess is supplemented by a functional or systemic value that is determined by the adequateness of a moral subject's 'fit' into its environment. The 'fit' or ecologically supervenient value will vary with circumstances but is a practicable way to determine how best, morally, to treat different organisms—and of necessity gives a prominent role to the *science* of ecology. This ecologically supervenient value also can take into account the moral subject's status as a moral agent, moral agency being relevant to ecological 'fit' because of the moral agent's capacity to act in the interests (or not) of others and so actively promote and preserve (or not) 'the integrity, stability, and beauty of the biotic community'.

However, adopting this supervenient value as a criterion for acting requires the moral agent to take, in part at least, what Habermas calls an 'objectivating' attitude to nature and natural things. Habermas says the only 'theoretically fruitful attitude' to nature is the 'objectivating attitude of the natural-scientific experimenting observer'[8] but this is a pragmatic position that does not rule out an approach predicated on the value-for-itself of organisms. The objectivating approach does not require that the objects of our observation should themselves be valueless or have only an instrumental value. The objectivating approach is an epistemic position: it is a position a conscious agent may take if he or she wishes to have some measure of control over his/her causal powers and the effects such powers have. It does not, however, preclude an agent's desire for control over causal powers to be itself prompted by a *feeling* of responsibility for anything which has a good of its own and hence whose interests can be harmed or benefitted by the agent's causal powers.

This book has attempted to demonstrate the feasibility of an ethic predicated on the unique value of living individuals—in particular, sentient individuals—as self-maintaining, self-realizing entities with a

8 Cited in C. F. Alford, Science and the Revenge of Nature: Marcuse and Habermas, Tampa and Gainesville: Universities of Florida Press, 1985, p. 147.

good of their own. In this ethic, moral agents are placed under an obligation to care for, to take responsibility for, the unique value possessed by each living individual. The obligation is a primary part of the being of moral agents—without the obligation to care for others the moral agent would not be who or what he or she is. The obligation falls on us as moral agent simply because of what we are.

References

Agar, N. *Life's Intrinsic Value: Science, Ethics and Nature*. New York: Columbia University Press, 2001.

Alford, C. F. *Science and the Revenge of Nature: Marcuse and Habermas*. Tampa and Gainesville: Universities of Florida Press, 1985.

Ackrill, J.L. 'Aristotle's Distinction Between *Energeia* and *Kinesis*.' Renford Bambrough ed. *New Essays on Plato and Aristotle*. New York: Humanities Press, 1965.

Aquinas, St. T. *The Summa Contra Gentiles of St. Thomas Aquinas*. Books I and II. Trans. English Dominican Fathers. London: Burns, Oates and Washbourne, 1924.

---. *Summa Contra Gentiles*. Book 2: Creation. Trans. James F. Anderson. Notre Dame: University of Notre Dame P, 1975.

---. *Commentary on Aristotle's Physics*. Book II. Trans. Richard Blackwell, Richard Spath and W. Edmund Thirlkel. London: Routledge and Kegan Paul, 1963.

---. *Summa Contra Gentiles*. Bk I. Trans. A. Pegis. Notre Dame: University of Notre Dame P, 1975.

---. *Summa Contra Gentiles*. Bk III. Trans. V.J. Bourke. 1956. University of Notre Dame P, 1975.

Aristotle. *De Anima*. Trans. D.W. Hamlyn. Oxford: Clarendon, 1968. (The Clarendon Aristotle).

---. *De Anima*. Trans. R.D. Hicks. Amsterdam: Adolf M. Hakkert, 1965.

---. *The Complete Works of Aristotle*. J. Barnes ed. Princeton: Princeton University Press/Bollingen, 1984.

---. *Generation of Animals*. Trans. A.L. Peck. Cambridge, Mass.: Harvard University Press, 1990. (Loeb edition).

---. *Metaphysics*. Trans. H. Tredennick. London: Wm. Heinemann, 1936. (Loeb edition).

---. *The Nicomachean Ethics*. Trans. H. Rackham. Cambridge, Mass.: Harvard University Press, 1990. (Loeb edition).

---. *The Nicomachean Ethics.* Trans. J.A.K. Thomson and H. Tredennick, Harmondsworth: Penguin, 1976.

---. *On Sense and Sensible Objects (De Sensu)*. Trans. W.S. Hett. Cambridge, Mass: Harvard University Press, 1986. (Loeb edition)

---. *On the Soul*. Trans. W.S. Hett. London: Heinemann, 1935. (Loeb edition).

---. *Physics.* Trans. P. Wicksteed and F. Cornford. London: Wm. Heinemann, 1929. (Loeb edition).

---. *Physics*. Trans. R.P. Hardie and R.K. Gaye. J. Barnes ed. *The Complete Works of Aristotle*.

Attfield, R. *The Ethics of Environmental Concern*. Oxford: Basil Blackwell, 1983.

Baier, A. 'Trust and antitrust'. *Ethics*. Vol. 96, No. 2 ,1986, Jan.

Bains, P. 'Umwelten'. *Semiotica.* Issue 134, 2001.

Bambrough, R. ed. *New Essays on Plato and Aristotle.* London: Routledge and Kegan Paul, 1965.

Barbieri, M. 'Has Biosemiotics Come of Age?'. Barbiero, M. ed. *Introduction to Biosemiotics*. Dordrecht: Springer, 2007.

Bateson, G. *Steps to an Ecology of Mind*. New York: Bantam, 1988.

Bauman, Z. *Postmodern Ethics*. Oxford: Blackwell, 1993.

Beckermann, A., H. Flohr and J. Kim eds. *Emergence or Reduction? Essays on the Prospects of Nonreductive Physicalism.* Berlin: Walter de Gruyter, 1992.

Beckermann, A. Introduction. *Emergence or Reduction?* Eds. Beckermann et al.

Beer, S. Preface. *Autopoiesis and Cognition*. By Maturana and Varela.

Bible, The Holy. Revised Standard Version. 2nd ed. London: Oxford University Press, 1971.

Blackburn, S. *The Oxford Dictionary of Philosophy*. Oxford: Oxford University Press, 1994.

Blair, G. *Energeia and Entelecheia: 'Act' in Aristotle*. Ottowa: University of Ottowa P, 1992.

Blyth, D. 'Universal Teleology in Aristotle.' unpub. ms.

---. 'Wholes, Parts, and Sequences in Aristotle.' *International Philosophical Quarterly* 34.4, No. 136 (December 1994): 453:463.

Bookchin, M. 'Recovering Evolution: A Reply to Eckersley and Fox.' *Environmental Ethics* 12 (Fall 1990): 253-274.

Bowler, P.J. *The Fontana History of the Environmental Sciences*. London: Fontana/HarperCollins, 1992

Brennan, A. *Thinking About Nature: An Investigation of Nature, Value and Ecology*. London: Routledge, 1988.

Brennan, R. *Thomistic Psychology: a Philosophic Analysis of the Nature of Man*. New York: Macmillan, 1941.

Brier, S. 'Foreword: Sebeok's biosemiotic contribution to Cybersemiotics'. *Cybernetics and Human Knowing*. Vol. 10, No. 1, 2002.

Burger, R. 'Is Each Thing The Same as its Essence?: On *Metaphysics* Z. 6-11.' *Review of Metaphysics* 41 (September 1987): 53-76.

Callicott, J. B. 'Intrinsic Value, Quantum Theory, and Environmental Ethics.' *Environmental Ethics* 7 (1985).

---. 'The Search for an Environmental Ethic'. *Matters of Life and Death*. 2nd ed. Tom Regan ed. New York: Random House, 1986.

Capra, F. *The Tao of Physics*. London: Fontana/Flamingo, 1983.

Chantraine, P. *Dictionnaire Etymologique de la Langue Grècque: Histoire des Mots*. Paris: Editions Klincksieck, 1968. Vol. 3.

Chebanov, S. 'Umwelt as life world of a living being'. *Semiotica*, Issue 134, 2001.

Cicero [Marcus Tullius Ciceronis]. *De Natura Deorum.* Libri Tres with commentary by J.B. Mayor. Cambridge: Cambridge University Press, 1883 (Mayor edition).

---. *De Natura Deorum.* Libri 2 & 3. Arthur Stanley Pease ed. Cambridge, Mass: Harvard University Press, 1958. (Pease edition).

---. *De Natura Deorum.* W. Ax ed. Stuttgart: B.G. Teubner, 1961. (Teubner edition) *Scripta Quae Manserunt Omnia* Fasc. 45.

---. *De Finibus Bonorum et Malorum.* http://www.epicurus.net/en/finibus.html. Accessed March, 2011.

Clark, J. Introduction to Part 4, 'Social Ecology'. *Environmental Philosophy: From Animal Rights to Radical Ecology*. Michael Zimmerman et al. eds.

Clark, S.R.L. *The Moral Status of Animals*. Oxford: Oxford University Press, 1977.

---. *Animals and Their Moral Standing*. London: Routledge, 1997.

Collingwood, R.G. *The Idea of Nature*. Oxford: Clarendon, 1945.

Cooper, J.M. 'Aristotle on Natural Teleology'. *Language and Logos*. M. Schofield and M. Nussbaum eds. Cambridge: Cambridge University Press, 1982.

Curley, E. 'Spinoza's Metaphysics: An Interpretation.' Diss. Duke U, 1963. Ann Arbor: UMI, 1964. 644732.

Deleuze, G. *Expressionism in Philosophy: Spinoza*. Trans. Martin Joughin. New York: Zone Books, 1992.

Devall, B. and G. Sessions, eds. *Deep Ecology: Living as if Nature Mattered*. Salt Lake City: Gibbs Smith, 1985.

Diogenes Laertius. *De Vitis, Dogmatibus et Apophthegmatibus Clarorum Philosophorum*. Book Z, 85. www.mikrosapoplous.gr/dl/dl07.html #Zinon. Accessed Nov. 2008.

Donagan, A. 'Morality, Property and Slavery'. *The Philosophical Papers of Alan Donagan*. J.E. Malpas ed. Chicago and London: University of Chicago Press, 1994. Vol. 2, pp.107-131.

---. 'Philosophical Ethics: What It Is and What It Should Be'. *The Philosophical Papers of Alan Donagan*. J.E. Malpas ed. Vol. 2, pp.262-282.

---. 'Spinoza's Dualism.' *The Philosophy of Baruch Spinoza*. R. Kennington ed. *Studies in Philosophy and the History of Philosophy* 7. Washington DC: Catholic University of America Press, 1980, pp.89-102.

---. 'The Moral Theory Almost Nobody Knows: Kant's'. *The Philosophical Papers of Alan Donagan*. J.E. Malpas ed. Vol. 2, pp.144-152.

---. 'The Relation of Moral Theory to Judgments: A Kantian Review'. *The Philosophical Papers of Alan Donagan*. J.E. Malpas ed. Vol. 2, pp.194-216.

Elton, C. *Animal Ecology*. Oxford: Oxford University Press, 1927, pp.63-4.
Feinberg, J. *Social Philosophy*. Englewood Cliffs, NJ: Prentice Hal, 1973.
---. 'The Rights of Animals and Unborn Generations.' *Philosophy and Environmental Crisis*. ed. W. Blackstone. Athens, Ga.: University of Georgia, 1974.
---. 'Abortion'. *Matters of Life and Death*. ed. Tom Regan. 2nd ed. New York: Random House, 1986.
Fleischaker, G. R. *Autopoiesis: System Logic and Origins of Life*. Diss. Boston U, 1988. Ann Arbor: UMI, 1988. 8814650.
---. 'Guest Editor's Introduction: The Question Concerning the Ontology of Autopoiesis and the Limits of its Utility.' *International Journal of General Systems* 21, 1992 pp.131-141.
Fox, W. *Toward a Transpersonal Ecology: Developing New Foundations for Environmentalism*. Boston: Shambhala, 1990.
Frey, R.G. *Interests and Rights: the Case Against Animals*. Oxford: Clarendon, 1980.
Frankfurt, H. *Necessity, Volition and Love*. Cambridge: Cambridge University Press, 1999.
Funk, I. ed. *A Standard Dictionary of the English Language*. New York: Funk and Wagnall, 1901.
Gardiner, R. A. and B. Gardiner. 'Teaching Sign-Language to a Chimpanzee.' *Science* 165, 1969, pp.664-672.
Gill, M. L. *Aristotle on Substance: the Paradox of Unity*. Princeton, NJ: Princeton University Press, 1989.
Glanville, R. 'The Same is Different.' *Autopoiesis: A Theory of Living Organization*. ed. Milan Zeleny.
Glare, P.G.H., ed. *The Oxford Latin Dictionary*. Oxford: Clarendon, 1982.
Goodpaster, K. 'On Being Morally Considerable.' Michael Zimmerman et al. eds. *Environmental Philosophy*. Englewood Cliffs, NJ: Prentice Hall, 1993.
Gotthelf, A. 'Aristotle's Conception of Final Causality.' *The Review of Metaphysics* 30, 1976, pp.226-254.

---. 'Introduction'. *Aristotle on Nature and Living Things.* Allan Gotthelf ed. Pittsburgh, Pa: Mathesis, 1985; Bristol: Bristol Classical Press, 1985.
Greisch, J. 'Otherness: an ethical category?' G. de Stexhe, and J. Verstraeten, eds. *Matters of Breath: foundations for professional ethics*. Leuven: Peeters, 2000.
Grene, M. *The Understanding of Nature.* Dordrecht: D. Reidel, 1974.
---. *The Knower and the Known*. London: Faber, 1966.
---. 'Aristotle and Modern Biology.' *Topics in the Philosophy of Biology*. M. Grene and E. Mendelsohn eds. Dordrecht: Reidel, 1976, pp.3-36.
Grene, M. and E. Mendelsohn, eds. *Topics in the Philosophy of Biology*. Dordrecht: Reidel, 1976.
Gruen, L. 'The Moral Status of Animals', *The Stanford Encyclopedia of Philosophy* (Fall 2010 Edition), Edward N. Zalta (ed.), http://plato.stanford.edu/archives/fall2010/entries/moral-animal.
Habermas, J. *Moral Consciousness and Communicative Action.* Trans. C. Lenhardt and S. Weber Nicholsen. Cambridge: Polity, 1990.
Hamblet, W. 'To being or not to being'. Tymieniecka, A. ed. *Analecta Husserliana* LXXXIV, 2005.
Hampshire, S. *Spinoza*. London: Faber and Faber, 1956.
Harré, R. and E.H. Madden. *Causal Powers*. Oxford: Basil Blackwell, 1975.
Heidegger, M. *An Introduction to Metaphysics*. Trans. Ralph Manheim. New Haven: Yale University Press, 1959.
---. 'On the Being and Conception of PHYSIS in Aristotle's *Physics* B. 1.' Trans. Thomas J. Sheehan. *Man World* 9, 1976.
---. *On The Essence Of Language: The Metaphysics of Language and the Essencing of the Word; Concerning Herder's Treatise On the Origin of Language*, Trans., Gregory Torres, W. and Y. Unna, New York: SUNY Press, 2004.
Hoffmeyer, J. 'Biosemiotics and Ethics'. N. Witoszek, and E. Gulbrandsen, eds. *Culture and Environment: Interdisciplinary Approaches.* Oslo: Centre for Development and the Environment, 1993.

Hobbes, T. *Leviathan.* (1651). London: Penguin, 1985.
Holy Bible, The. Revised Standard Version. 2nd. ed. London: Oxford University Press, 1971.
James, S. 'Human Virtues and Natural Values'. *Environmental Ethics* 28, 2006.
Johnson, L. E. *A Morally Deep World: an essay on moral significance and environmental ethics.* Cambridge: Cambridge University Press, 1991.
Jonas, H. *The Imperative of Responsibility: In Search of an Ethics for the Technological Age.* Chicago: University of Chicago Press, 1984.
Jones, G. et al., eds. *Dictionary of Environmental Science.* Glasgow: Collins, 1990.
Kant, I. *Critique of Judgment.* Trans. W.S. Pluhar. Indianapolis: Hackett, 1987.
---. *Critique of Pure Reason.* Trans. N. Kemp Smith. London: Macmillan, 1992.
---. *Groundwork of the Metaphysic of Morals.* 3rd ed. Trans. H.J. Paton. New York: Harper, 1964.
---. *The Metaphysic of Morals.* Trans. M. McGregor. Cambridge: Cambridge University Press, 1991.
Kenny, V. 'On the Subject of Autopoiesis and Its Boundaries: Does the Subject Matter?' *International Journal of General Systems* 21, 1992, pp.187-196.
Kim, J. *Supervenience and Mind: Selected Philosophical Essays.* Cambridge: Cambridge University Press, 1993.
Kleinig, J. 'Crime and the Concept of Harm.' *American Philosophical Quarterly* 15.1, January 1978.
Kosman, L.A. 'Aristotle's Definition of Motion.' *Phronesis* 14, 1969, pp.40-62.
Krampen, M. 'No Plant —No Breath'. *Semiotica,* 134:1/4, 2001.
Kristeller, P. O. 'Stoic and Neoplatonic Sources of Spinoza's Ethics'. *History of European Ideas* Vol. 5, 1984.

Lachterman, D. R. 'The Physics of Spinoza's ETHICS.' *Spinoza: New Perspectives*. R.J. Shahan and J.I. Biro eds. Norman: University of Oklahoma, 1978, pp.71-112.

Lawson-Tancred, H. 'Aristotle'. *Philosophy: A Guide Through the Subject*. A.C. Grayling ed. Oxford: Oxford University Press, 1995, pp. 398-440.

Leopold, A. *A Sand County Almanac and Sketches Here and There*. (1949). New York: Oxford University Press, 1987.

Lévinas, E. *Ethics and Infinity: Conversations with Philippe Nemo*. Trans. Richard Cohen. Pittsburgh: Duquesne University Press, 1985.

Levy, Z. *Baruch or Benedict, On some Jewish Aspects of Spinoza's Philosophy*. New York: Peter Lang, 1989.

Lewis, D.K. *On the Plurality of Worlds*. Oxford: Basil Blackwell, 1986.

Llewellyn, J. *The Middle Voice of Ecological Conscience*. London: Macmillan, 1991.

---. *Emmanuel Lévinas: the Genealogy of Ethics*. London: Routledge, 1995.

Liddell, H. G. and R. Scott, *A Greek-English Lexicon*. 9th ed. Revised H. S. Jones and R. McKenzie. Oxford: Clarendon, 1990.

Little, W., H.W. Fowler and J. Coulson eds. *The Shorter Oxford English Dictionary on Historical Principles*. C. Onions rev. & ed. 3rd ed. Oxford: Clarendon, 1985.

Lovelock, J. E. *Gaia: a New Look at Life on Earth*. Oxford: Oxford University Press, 1979.

McLaughlin, B. 'Varieties of Supervenience.' E. Savellos and Ü.Yalçin eds. *Supervenience: New Essays*.

Macpherson, C.B. 'Introduction'. Hobbes, T. *Leviathan*. London: Penguin, 1985.

Mathews, F. *The Ecological Self*. London: Routledge, 1991.

Maturana, H. and F. Varela. *Autopoiesis and Cognition: The Realization of The Living*. Boston/Dordrecht: Reidel, 1980.

---. *The Tree of Knowledge: The Biological Roots of Human Understanding*. Boston: Shambhala, 1992.

---. 'Autopoiesis: Reproduction, Heredity and Evolution.' *Autopoiesis, Dissipative Structures, and Spontaneous Social Orders.* M. Zeleny ed.

---. 'The Organization of the Living: A Theory of the Living Organization.' *International Journal of Man-Machine Studies* Vol. 7, 1975 pp.313-332.

Merleau-Ponty, M. *Nature: Course Notes From The College de France.* Compiled D. Seglard, Trans. R. Vallier. Evanston: Northwestern University Press, 2003.

Merchant, C. *The Death of Nature: Women, ecology and scientific revolution.* New York: Harper and Row, 1980.

Mill, J.S. 'Utilitarianism.' *John Stuart Mill and Jeremy Bentham: Utilitarianism and Other Essays.* A. Ryan ed. Harmondsworth: Penguin, 1987.

Millett, S. *Autopoiesis and immanent teleology: toward an Aristotelian environmental ethic.* Diss. Perth: Murdoch University, 1997.

Moore, G. E. 'The Conception of Intrinsic Value.' *Principia Ethica.* Cambridge: Cambridge University Press, 1993.

Mourelatos, A. P. 'Aristotle's 'Powers' and Modern Empiricism.' *Ratio* 9.2, 1967, pp.97-104.

Murray, J.A.H. ed. *The Oxford English Dictionary.* Oxford: Clarendon, 1933.

Naess, A. '*Self*-realization: An Ecological Approach to Being in the World.' J. Seed et al eds. *Thinking Like a Mountain: Towards a Council of All Beings.* Philadelphia/Santa Cruz: New Society Publishers, 1988.

Nash, R. *The Rights of Nature.* Leichhardt, NSW: Primavera Press, 1990.

O'Neill, J. 1992. 'The Varieties of Intrinsic Value'. A. Brennan ed. *The Ethics of the Environment.* Aldershot: Dartmouth, 1995, pp.55-73.

Onions, C. ed. *The Oxford Dictionary of English Etymology.* Oxford: Clarendon, 1991.

Owen, G.E.L. 'Aristotle on the Snares of Ontology.' R. Bambrough ed. *New Essays on Plato and Aristotle.*
Owens, J. *The Doctrine of Being in the Aristotelian 'Metaphysics'*. 3rd ed. Toronto: Pontifical Institute of Mediaeval Studies, 1978.
Papy, J. 'Justus Lipsius'. *The Stanford Encyclopedia of Philosophy* (Autumn 2004 edition). Edward N. Zalta (ed.), URL = http://plato.stanford.edu/entries/justus-lipsius/#Rel
Passmore, J. *Man's Responsibility for Nature.* London: Gerald Duckworth, 1974.
Parkes, G. 'Human/Nature in Nietzsche and Taoism.' J.Baird Callicott and R. Ames eds.. *Nature in Asian Traditions of Thought.* New York: SUNY Press, 1989.
Plato. *Symposium.* Trans. W. Hamilton. Harmondsworth: Penguin, 1951.
---. *Gorgias.* Trans. W. Hamilton. Harmondsworth: Penguin. 1960.
---. *Phaedo*. Trans. B. Jowett. http://classics.mit.edu/Plato/phaedo.html.
Plotinus. *Psychic and Physical Treatises; comprising the second and third Enneads.* Trans. S. Mackenna. London: Philip Lee Warner, 1921.
Pluhar, W. 'Translator's Introduction'. Kant, I. *Critique of Judgment.* Indianapolis: Hackett, 1987.
Plumwood, V. *Feminism and the Mastery of Nature.* London: Routledge, 1993.
Regan, T. 'The Nature and Possibility of an Environmental Ethic.' *Environmental Ethics* 3, 1981.
---. *The Case for Animal Rights.* Los Angeles: University of California Press, 1983.
Reich, W. 2006. 'The Duty of Care: Implications for the professions'. Paper presented at the Australasian Association for Professional and Applied Ethics 13th annual conference. Sydney, June.
Ristau, C. ed. *Cognitive Ethology. The Minds of Animals: Essays in Honor of Donald R. Griffin.* Mahwah, NJ: Lawrence Erlbaum Associates, 1991.
Rodd, R. *Biology, Ethics and Animals.* Oxford: Clarendon, 1990.

Rollin, B. E. *Animal Rights and Human Morality.* New York: Prometheus, 1981.

---. 'The Legal and Moral Bases of Animal Rights'. H.B. Miller and W.H. Williams eds. *Ethics and Animals.* Clifton, NJ: Humana Press, 1983.

Rolston, H, III. 'Review of Andrew Brennan's *Thinking About Nature.*' *Environmental Ethics* 11 (Fall) 1989, pp.259-267.

---. *Environmental Ethics: Duties to and Values in the Natural World.* Philadelphia: Temple University Press, 1988.

---. 'Duties to Ecosystems.' J. Baird Callicott ed. *Companion to 'A Sand County Almanac'.* Madison: University of Wisconsin Press, 1987.

Sahtouris, E. *Gaia: The Human Journey From Chaos to Cosmos*. New York: Simon and Schuster, 1989.

Sandler, R. and P. Cafaro, eds. *Environmental Virtue Ethics*. Lanham MD: Rowman and Littlefield, 2005.

Sapontzis, S. F. 'The Moral Significance of Interests.' *Environmental Ethics* 4 (Winter) 1982, pp.345-358.

Saunders, J. L. *Greek and Roman Philosophy After Aristotle*. New York: Free Press-Macmillan, 1966.

Savellos, E. E. and Ü. D. Yalçin, eds. *Supervenience: New Essays.* Cambridge: Cambridge University Press, 1995.

Sebeok, T. 'The Sign Science and the Life Science'. *Applied Semiotics*. 3:6/7, 1999.

Sedley, D. 'Is Aristotle's Teleology Anthropocentric?' *Phronesis* 36.2, 1991, pp.179-196.

Sellars, W. *Empiricism and the Philosophy of Mind.* 1956. Cambridge: Harvard University Press, 1997.

Shoemaker, S. *Identity, Cause and Mind.* Cambridge: Cambridge University Press, 1984.

Simpson, D.P. ed. *Cassell's New Latin-English English-Latin Dictionary*. London: Cassell, 1959.

Simpson, J.A. and E.S.C. Weiner eds. *The Oxford English Dictionary*. 2nd ed. Oxford: Clarendon, 1989.

Singer, P. *Animal Liberation.* 2nd ed. New York: Harper Collins, 1990.

Smith, W. ed. *A Smaller Latin-English Dictionary.* London: John Murray, 1879.

Snyder, G. 'Buddhism and the Possibilities of a Planetary Culture.' B. Devall and G. Sessions eds. *Deep Ecology.*

Spinoza, Benedict de. *A Spinoza Reader: The 'Ethics' and Other Works*. Trans. and ed. Edwin Curley. Princeton: Princeton University Press, 1994.

---. *The Chief Works of Benedict de Spinoza.* Trans. R.H.M. Elwes. 1883. New York: Dover, 1955.

---. *Cogitata Metaphysica.* Trans. Edwin Curley. *The Collected Works of Spinoza.*

---. 'Korte Verhandlung/ Short Treatise'. *The Collected Works of Spinoza..*

Spinoza, Benedictus [Baruch] de. *The Collected Works of Spinoza.* Trans. and ed. Edwin Curley. Princeton: Princeton University Press, 1985.

Spinoza, Benedictus de. *Ethics*. Trans. Andrew Boyle ed. and rev. trans. G.H.R Parkinson. London: Everyman, 1989.

Stoffell, B. 'Commonwealth and Civility: A Study of Thomas Hobbes.' Diss. University of British Columbia, 1984.

Swanton, C. 'Heideggerian Environmental Virtue Ethics'. *Journal of Agricultural and Environmental Ethics*. Published online, 17 June 2009.

Taylor, C. *Sources of the Self: the making of modern identity.* Cambridge, Mass.: Harvard University Press, 1989.

Taylor, P. *Respect for Nature*. Princeton: Princeton University Press, 1986.

Tønnessen, M. 'Umwelt ethics.' *Sign Systems Studies* 31.1, 2003.

Uexküll, J. Von. *Umwelt und Innenwelt der Tiere*. Berlin: J. Springer, 1909.

Upshall, M. ed. *The Hutchinson Paperback Dictionary of Biography.* London: Random Century, 1990.

Varela, F., H. Maturana and R. Uribe. 'Autopoiesis: The Organization of Living Systems, its Characterization and a Model.' *Biosystems* 5, 1974, pp.187-196.
Vogel, L. 'The outcry of mute things: Hans Jonas's imperative of responsibility'. D. Macaulcy ed. *Minding Nature: the Philosophers of Ecology*. New York: The Guilford Press, 1996.
Vogel, L. 'Jewish Philosophies after Heidegger'. W. Davis, ed. *Taking Responsibility: Comparative Perspectives.* Charlottesville: University Press of Virginia, 2001.
Vogel, L. 'Overcoming Heidegger's Nihilism'. S. Fleischaker ed. *Heidegger's Jewish Followers*. Pittsburgh: Duquesne University Press, 2008.
Wall, D. *Green History*. London: Routledge, 1994.
Wardy, R. 'Aristotelian Rainfall or the Lore of Averages'. *Phronesis* 38.1, 1993, pp.18-28.
Weber, B.L. & I.E. Woodrow. *Functional Plant Biology*. 'Cassowary frugivory, seed defleshing and fruit fly infestation influence the transition from seed to seedling in the rare Australian rainforest tree, Ryparosa sp. nov. 1 (Achariaceae).' 31: 505-516. en.wikipedia.org/wiki/Cassowary accessed June 25, 2009.
Wheeler, J.A. 'Curved Empty Space Time As the Building Material of the Physical World: An Assessment'. E. Nagel et al. eds. *Logic, Methodology and Philosophy of Science*. Stanford: Stanford University Press, 1962.
Williams, M. 'Rights, Interests and Moral Equality.' *Environmental Ethics* 2, 1980, pp.149-161.
Wilson, E.O. *Biophilia*. Harvard: Harvard University Press, 1984.
Wolfson, H.A. *The Philosophy of Spinoza: Unfolding the Latent Processes of His Reasoning.* 1934. New York: Schocken, 1969. Two volumes.
Woodfield, A. *Teleology*. Cambridge: Cambridge University Press, 1976.
Yovel, Y. *Spinoza and Other Heretics*. Vol. 1. *The Marrano of Reason*. Princeton: Princeton University Press, 1989.

---. *Spinoza and Other Heretics*. Vol. 2. *The Adventures of Immanence*. Princeton: Princeton University Press, 1989.
Zeleny, M. *Autopoiesis: A Theory of Living Organization*. New York: Elsevier North Holland, 1981.
---. ed. *Autopoiesis, Dissipative Structures, and Spontaneous Social Orders*. AAAS Selected Symposium No. 55. Boulder, Colo.: Westview Press, 1980.
Zeller, E. *Outlines of the History of Greek Philosophy*. 1931. 13th ed. Rev. W. Nestle. Trans. L.R. Palmer. New York: Dover, 1980.
Zimmerman, M. et al. eds. *Environmental Philosophy*. Englewood Cliffs, NJ: Prentice Hall, 1993.

Subject Index

Name Index

Berner Reihe philosophischer Studien

Band 1: Gerhard Heinzmann
Schematisierte Strukturen. Eine Untersuchung über den Idoneismus Ferdinand Gonseths auf dem Hintergrund eines konstruktivistischen Ansatzes. 185 S., 1982.

Band 2: Vincent F. Brunner
Probleme der Kausalerklärung menschlichen Handelns. 112 S., 1983.

Band 3: Jean-Claude Wolf
Sprachanalyse und Ethik. Eine Kritik der Methode und einiger Folgeprobleme sowie der Anwendung des universalen Präskriptivismus von Richard Mervyn Hare. 142 S., 1983.

Band 4: Duen Jau Marti-Huang
Die Gegenstandstheorie von Alexius Meinong als Ansatz zu einer ontologisch neutralen Logik. 136 S., 1984.

Band 5: Claudia Risch
Die Identität des Kunstwerks. VIII, 189 S., 1986.

Band 6: Andreas Graeser (ed.)
Mathematics and Metaphysics in Aristotle/Mathematik und Metaphysik bei Aristoteles. 333 S., 1987.

Band 7: Stephan Hottinger
Nelson Goodmans Nominalismus und Methodologie. 94 S., 1988.

Band 8: Michael Schrijvers
Spinozas Affektenlehre. 233 S., 1989.

Band 9: Andreas Gunkel
Spontaneität und moralische Autonomie. Kants Philosophie der Freiheit. IX, 238 S., 1989.

Band 10: Andreas Bächli
Untersuchungen zur pyrrhonischen Skepsis. XI, 90 S., 1990.

Band 11: Hansueli Flückiger
Sextus Empiricus. Grundriss der pyrrhonischen Skepsis. Buch I – Selektiver Kommentar. 126 S., 1990.

Band 12: Gérard Bornet
Naive Semantik und Realismus. Sprachphilosophische Untersuchung der Frühphilosophie von Bertrand Russell (1903–04). 315 S., 1991.

Band 13: Herbert Schweizer
Bedeutung. Grundzüge einer internalistischen Semantik. 201 S., 1991.

Band 14: Jürg Freudiger
Kants Begründung der praktischen Philosophie: systematische Stellung, Methode und Argumentationsstruktur der „Grundlegung zur Metaphysik der Sitten". 123 S., 1993.

Band 15: Erwin Sonderegger
Aristoteles, Metaphysik Z 1-12: philosophische und philologische Erwägungen zum Text. VIII, 369 S., 1993

Band 16: Lutz Danneberg / Andreas Graeser / Klaus Petrus (Hg.)
Metapher und Innovation: die Rolle der Metapher im Wandel von Sprache und Wissenschaft. 324 S., 1995.

Band 17: Gérard Bornet
Die Bedeutung von „Sinn" und der Sinn von „Bedeutung": auf dem Weg zu einem gemeinsprachlichen Wörterbuch für formale Philosophie. XVI, 246 S., 1996.

Band 18: Cinzia Ferrini
Guida al „De orbitis planetarum" di Hegel ed alle sue edizioni e traduzioni. 259 S., 1995

Band 19: Urs Bruderer
Verstehen ohne Sprache. Zu Donald Davidsons Szenario der radikalen Interpretation. 80 S., 1997.

Band 20: Peter Zimmermann
Soziologie als Erkenntniskritik. Die Wissenssoziologie Karl Mannheims. 64 S., 1998.

Band 21: Thomas Gfeller
Was ist wichtig: Beschreibung, Wertung und ethische Theorie. 112 S., 1998.

Band 22: Gerald Bechtle
The Anonymous Commentary on Plato's „Parmenides". 285 S., 1999.

Band 23: Ursula Thomet
Kunstwerk, Kunstwelt, Weltsicht. Arthur C. Dantos Philosophie der Kunst und Kunstgeschichte. 112 S., 1999.

Band 24: Pierfrancesco Basile
Experience and Relations. An Examination of F.H. Bradley's Conception of Reality. 197 S., 1999.

Band 25: Andreas Graeser
Prolegomena zu einer Interpretation des zweiten Teils des Platonischen Parmenides. 60 S., 1999

Band 26: Martin Flügel / Thomas Gfeller / Charlotte Walser
Werte und Fakten. Eine Dichotomie im Spiegel philosophischer Kontroversen.
268 S., 1999.

Band 27: Rebecca Iseli
Kants Philosophie der Mathematik. Rekonstruktion – Kritik – Verteidigung.
110 S., 2001.

Band 28: Helen Plüss
Bedeutungstheorie und philosophische Psychologie bei Paul Grice.
76 S., 2001.

Band 29: Alain Metry
Speusippos. Zahl – Erkenntnis – Sein. 198 S., 2002.

Band 30: Heiko Holweg
Methodologie der qualitativen Sozialforschung. Eine Kritik. 176 S., 2005.

Band 31: Thomas Ruprecht
Die Unbestimmtheit der Verursachung. Ein philosophischer Essay über Kausalität. 138 S., 2003.

Band 32: Evangelos Zoidis
Notes on a Metaphysics of Presence. 113 S., 2003.

Band 33: Philipp Keller
Ein Dritter Blick auf Descartes' Erste Philosophie. Eine analytische Interpretation der cartesianischen Metaphysik.
In Vorbereitung.

Band 34: Marcel van Ackeren
Heraklit. Vielfalt und Einheit seiner Philosophie. 167 S., 2006.
ISBN 978-3-03910-815-2.

Band 35: Mika Ojakangas
A Philosophy of Concrete Life. Carl Schmitt and the Political Thought of Late Modernity. 2nd revised edition, 225 S., 2006.
ISBN 978-3-03910-963-0 / US-ISBN 978-0-8204-8362-7.

Band 36: Erwin Sonderegger
Aristoteles' Metaphysik Λ: Ein spekulativer Entwurf. Einführung, Übersetzung, Kommentar. XXXII, 514 S., 2008. ISBN 978-3-03911-477-1.

Band 37: Thomas Schindler-Wunderlich
Kritik der neuzeitlichen Wunderkritik. Eine religionsphilosophische Studie.
217 S., 2008. ISBN 978-3-03911-510-5.

Band 38: Gianluigi Segalerba: Semantik und Ontologie. Drei Studien zu Aristoteles. In Vorbereitung.

Band 39: François Grandjean: Aristoteles' Theorie der praktischen Rationalität. 346 S., 2009. ISBN 978-3-03911-674-4

Band 40: Charlotte Walser: Personen – Inwiefern wir sind, wofür wir uns halten. 235 S., 2009. ISBN 978-3-0343-0080-3

Band 41: Paul Taborsky: The Logic of Cultures. Three Structures of Philosophical Thought. 266 S., 2010. ISBN 978-3-0343-0378-1

Band 42: Alba Papa-Grimaldi: Orphans of the One or the Deception of the Immanence. Essays on the Roots of Secularization. 239 S., 2010. ISBN 978-3-0343-0410-8

Band 43: Stefano Marino: Gadamer and the Limits of the Modern Techno-Scientific Civilization. 295 S., 2011. ISBN 978-3-0343-0663-8

Band 44: Stephan Millett: Aristotle's Powers and Responsibility for Nature. 303 S., 2011. ISBN 978-3-0343-0679-9

Bände 1–32 erschienen im Paul Haupt Verlag, Bern